STO

ACPL ITEM
DISCARDED

S0-BUX-351

Crossing
The Social
Man
Corbett, J. Martin, 1956-
Crossing the border

DO NOT REMOVE
CARDS FROM POCKET

ALLEN COUNTY PUBLIC LIBRARY
FORT WAYNE, INDIANA 46802

You may return this book to any agency, branch,
or bookmobile of the Allen County Public Library.

DEMCO

The Springer Series on

ARTIFICIAL INTELLIGENCE AND SOCIETY

Series Editor: KARAMJIT S. GILL

Knowledge, Skill and Artificial Intelligence
Bo Göranzon and Ingela Josefson (Eds.)

Artificial Intelligence, Culture and Language:
 On Education and Work
Bo Göranzon and Magnus Florin (Eds.)

Designing Human-centred Technology:
 A Cross-disciplinary Project in Computer-aided Manufacturing
H. H. Rosenbrock (Ed.)

The Shape of Future Technology:
 The Anthropocentric Alternative
Peter Brödner

Artificial Intelligence and Human Institutions
Richard Ennals

J. Martin Corbett, Lauge Baungaard Rasmussen
and Felix Rauner

Crossing the Border

The Social and Engineering Design of Computer Integrated Manufacturing Systems

With 27 Figures

Springer-Verlag
London Berlin Heidelberg New York
Paris Tokyo Hong Kong

J. Martin Corbett, MA, AFBPsS
Warwick Business School, University of Warwick,
Coventry CV4 7AL, UK

Lauge Baungaard Rasmussen, MA
Institute of Social Science, Danish Technical University,
DK-2800, Lyngby, Denmark

Felix Rauner, MA, PhD
Department of Vocational Education, Institute of Technology and
Education, University of Bremen, 2800 Bremen 33, FRG

ISBN 3-540-19613-7 Springer-Verlag Berlin Heidelberg New York
ISBN 0-387-19613-7 Springer-Verlag New York Berlin Heidelberg

British Library Cataloguing in Publication Data
Corbett, J. Martin 1956–
Crossing the border: the social and engineering design of computer integrated
  manufacturing systems.
  1. Manufacture. Applications of computer systems
  I. Title II. Rasmussen, Lauge Baungaard 1947– III. Rauner, Felix 1941–
670.285
ISBN 3-540-19613-7

Library of Congress Cataloging-in-Publication Data
Corbett, J. Martin, 1956–
Crossing the border: the social and engineering design of computer integrated
  manufacturing systems/J. Martin Corbett, Lauge Baungaard Rasmussen, and Felix
  Rauner.
  p. cm. – (The Springer series on artificial intelligence and society)
  Includes bibliographical references.
  ISBN 0-387-19613-7 (U.S.)
  1. Computer integrated manufacturing systems – Design and construction.
  2. Human engineering. I. Rasmussen, Lauge Baungaard. 1947–. II. Rauner,
  Felix. III. Title. IV. Series.
TS155.6.C68 1991   90–10082
670'.285–dc20   CIP

Apart from any fair dealing for the purposes of research or private study, or criticism
or review, as permitted under the Copyright, Designs and Patents Act 1988, this
publication may only be reproduced, stored or transmitted, in any form or by any
means, with the prior permission in writing of the publishers, or in the case of
reprographic reproduction in accordance with the terms of licences issued by the
Copyright Licensing Agency. Enquiries concerning reproduction outside those
terms should be sent to the publishers.

© Springer-Verlag London Limited 1991
Printed in Great Britain

Typeset by Nuts and Muttons Typesetting Ltd, Linton, Cambridgeshire
Printed and bound by Alden Press Ltd, Osney Mead, Oxford
2128/3916-543210 Printed on acid-free paper

Foreword

The gap between the potential of technology and its reality has become a gaping chasm. It was in an attempt to bridge that chasm that Lucas workers embarked upon their extraordinary plan for socially useful production in the early 70s. The plan assumed international significance and its main proposals are now well known (Cooley 1987). One of its great strengths was that those involved began to think of themselves in their dual role in society: as producers and as consumers. They were therefore concerned at what they made, why they made it, and even questioned if they should make it at all. But they were also concerned about how they made things, e.g. the means of production. In the course of these discussions they initially suggested telechiric devices which "would respond to human knowledge and intelligence but would not objectivize it".

In the highly creative discussions which ensued, there emerged the concept of human-centredness. To the best of my knowledge it was used for the first time in the English language in February 1976 (by myself) at the press conference to launch the Lucas workers' plan for socially useful production. It was one strand which subsequently converged with two independent strands in Denmark and in Germany to provide a paradigmatic shift in designers' thinking; a shift which sought to deal with the problem of reducing human beings to abject machine appendages and thereby losing one of society's most precious assets which is the skill, ingenuity, creativity and intentionality of its people.

The background to the convergence of these three traditions is vividly posed in the book. It may be said to be a European approach in the sense that it contrasts starkly with the Tayloristic form of production organization in the United States. It places the individual human being with his or her competence and motivations as a central tenet, but views the individual not in isolation but as part of a creative interaction with fellow human beings in a collaborative model of production. In doing so, it was able to draw on the rich variety of different traditions in Europe, and the resultant systems may be said to be those which can accord with different cultural and educational backgrounds and experiences across Europe.

A recent report highlights the importance of this diversity which it identifies as one of Europe's key strengths rather than weaknesses, and it advocates policies which will enhance these diversities rather than see Europe from 1992 decompose into the sort of melting pot one saw in the United States at the turn of the century (see the publication *European Competitiveness in the 21st Century*). Thus we may be seeing the emergence of European forms of manufacturing technology in an enlarged Europe which has a geographical area large and diverse enough, a market dynamic enough, an economy powerful enough and a cultural base rich and diverse enough to develop appropriate human-centred forms of technology rather than accept the notion of the one best way at the macro level.

The concept of human-centred systems is by no means problem free as the authors freely admit – indeed a strength of the book is that it describes vividly the historical, epistemological, technological and organizational issues which gave rise to the idea in the first instance. They show it to be as yet a delicate plant which if uprooted from the fertile soil within which it germinated could well deform into a rather worthless variant of that other anaemic plant, "user friendliness". A full blooming of its potential is conditional upon its being located in an appropriate context. As such the book challenges much of the given wisdom and correctly cautions the reader that information technology has grown up in an environment of mathematical and natural sciences and has therefore been dominated by the positive epistemology of practice. This practice rests on three dichotomies: the separation of means and end, the separation of research from practice and the separation of knowing from doing.

The challenge is not just at the level of the technical aspects of the design although these are fascinating. The book highlights and seeks to answer some of the problems involved in the organization and operation of multidisciplinary design teams. As the instigator of ESPRIT 1217, I was deeply conscious of the need to take a multidisciplinary approach to the design problem but was also aware that this had not been done previously and that tensions would be inevitable. It was evident from the outset that the engineers and the social scientists were using different cognitive maps. Furthermore, given that the project was to be completed in three years, and understanding the lead time in ordering and procuring the manufacturing hardware, the engineers felt pressed to make decisions very early on. In doing so, they were however closing off important options. The social scientists on the other hand were dealing with more timeless categories such as alienation and subordination. The tensions this created bordered on occasion in open conflict. Only the goodwill of all involved, and a shared intention to improve the quality of working life, transcended professional differences and provided for the rich synthesis which was the final outcome.

But there were real difficulties en route. The book describes methods of overcoming these difficulties and of creating collaborative frameworks in which the different traditional approaches are transformed by interactions with the others. It would be difficult to overstate the importance of this process or indeed even to suggest that the project got beyond reaching even partial solutions in this key area of human and professional interaction in multidisciplinary design teams. It was precisely the crossing of these boundaries which made the project so challenging and exciting and also provided the title for this book.

The book is a controversial one. Even with the hindsight of experience it offers no glib solutions, but rather highlights issues which others embarking on similar design programmes can take into account. That experience would not have been possible without the funding from the EEC for this £5 million project which now looks set to produce forms of manufacturing technology which accord much more closely with the reality of European industry which is based on small and medium-sized enterprises with highly skilled and flexible workforces.

Although the issues raised here are within the context of manufacturing technology, I hold that they are in fact more universally applicable. At a time when there are models of universities as factories in which students are referred to as commodities, examinations as quality control procedures, graduation as delivery and the professors as operators (Cooley 1987), it behoves all professions to look carefully at the mechanistic concepts which are now beginning to dominate many professional areas. As new technology in the form of CAD, AI and expert systems is increasingly deployed, the real point at issue is if the systems will be computer-aided design or computer-controlled design and whether they are expert systems or expert replacement systems. Medicine, law and architecture, to mention but a few, are now confronted with these issues as AI and expert systems are increasingly contemplated for these professional domains.

Will the systems be designed as tools in the same sense in which Heidegger conceptualized it (Ehn 1988), or will they be machines which reduce human beings to abject appendage status? These are burning issues facing industrial society as we approach the year 2000. Indeed, these are some of the issues which will define our humanity and determine the future of our species in the coming millenium. The twenty-first century is a mere ten years away and its coming provides a powerful and psychological stimulus to re-examine the values, objectives and needs of our industrial society. It will constitute the end of an extraordinary millenium in which we have seen the decline of feudalism, the growth of our cities, industrial society and, above all, the development of that double-edged weapon, Western science and technology. It has been a journey which displayed the great strengths and the weaknesses of modern science and technology in which the delinquent genius of our species has produced both the beauty of Venice and the hideousness of Chernobyl, the caring, diagnostic potential of the

X-ray and the awesome destructiveness of the neutron bomb.

This book vividly invites us to build on that which is creative and humanistic and caring in our manufacturing technology. It will be an important contribution to a profound debate which hopefully will lay the basis for forms of manufacturing technology which are appropriate to our species as we approach the twenty-first century, a century in which we shall have to move from economy of scale to economy of scope *(European Competitiveness in the 21st Century)*. The book highlights three borders which have to be crossed. In doing so it implicitly identifies the greatest border of all – that in our own consciousness.

Mike Cooley

References

Cooley MJE (1987) Architect or Bee? Chatto and Windus, London

European competitiveness in the 21st century: the integration of work, culture and technology. A 95-page report available free from FAST, Rue de la Loi 200, B-1049 Brussels, Belgium.

Ehn P (1988) Work-oriented design of computer artefacts. Arbetslivscentrum, Stockholm

Contents

Introduction		1
1.	**The Work-Oriented Shaping Perspective: The Problem Setting and Perspectives**	**5**
1.1	Do We Need a New Paradigm?	5
1.2	The Technique-Oriented Approach	6
	1.2.1 Background and Assumptions	6
	1.2.2 Example	7
	1.2.3 "Weaknesses" in the Practical Implementation	8
1.3	The Sociotechnical Approach	9
	1.3.1 Background and Assumptions	9
	1.3.2 Examples	10
	1.3.3 "Weaknesses" in the Practical Implementation	11
1.4	The Human-Centred Approach	12
	1.4.1 Background and Assumptions	12
	1.4.2 Examples	13
	1.4.3 Possible Limitations and Challenges of the Human-Centred Approach	15
1.5	Towards a Work-Oriented Shaping Approach	15
1.6	Summary	19
2	**The Interdisciplinary Design of Computer-Integrated Manufacturing Systems: ESPRIT Project 1217 (1199) Part One**	**21**
2.1	Introduction	21
2.2	Experiences of Interdisciplinary Collaboration During the First Year of the Project	23
	2.2.1 The UK CAM Group	23
	2.2.2 The Danish CAD Group	26
	2.2.3. The German CAP Group	29
2.3	Evaluation of the First-Year Experiences in the Three National Groups	33
	2.3.1 National Differences	33
	2.3.2 Critical Evaluation of Interdisciplinary Collaboration During the First Year of the Project	34

2.4	The "Shaping Paper" – Intentions, Reactions and Influence	36
	2.4.1 Intentions: The International Social Science Group (ISSG)	36
	2.4.2 The Social Shaping of Technology and Work	37
A	Introduction	37
B	The Concept of Technological Shaping	38
B.1	Introduction	38
B.2	The Role of Technological Rationality	39
B.3	Technological Rationality and the Mechanistic World Picture	41
B.4	Work and Personality from a Holistic Perspective	42
B.4.1	The Search for a New Paradigm	42
B.4.2	The Concept of "Life-World" and "System"	43
B.4.3	The Concept of "Praxis"	44
B.4.4	The Relation Between Mind and Body	45
B.4.5	The Relationship between Working Conditions and the Personality	46
B.5	Fields of Technology Shaping	47
C	Dimensions of Human-Centred Work	50
C.1	Introduction	50
C.2	The Relationship of Work and Technology	51
C.3	Work and Communication	52
C.4	Work and Learning	54
C.4.1	The Relationship Between Cultivation (Bildung) and Qualification	55
D	Methods for Evaluating and Shaping Computer-Aided Technology and Work	57
D.1	Introduction	57
D.2	A Scenario	60
D.2.1	Range of Products	60
D.2.2	The Factory	60
D.2.3	The Production Island	62
D.2.4	The Workplace	63
D.3	The Scenario Related to the Dimensions and Criteria of Experiencing, Shaping and Evaluating Work	64
D.3.1	The Normative Perspective of the Shaping Process	64
D.3.2	Dimensions of Work	64
D.3.3	Example of How to Use the Dimensions of Work in the Shaping and Evaluation Process	69
D.3.4	Criteria for Human–Machine Interface Design	72
D.3.5	Summary	74
D.4	How to Shape and Evaluate Computer-Aided Technology and Work	75
D.4.1	Typical Shortcomings of Conventional Methods of System Development	75

	D.4.2 The Search after Methods in Accordance with the New Paradigm	75
	E References	77
	2.4.3 Reactions of the Engineers to the Shaping Paper	79
	2.4.4 The Influence of the Shaping Paper on the Project	80

3 Interdisciplinary Collaboration in the Second Half of ESPRIT Project 1217 (1199) 83

3.1 Experiences of Interdisciplinary Collaboration at the National Level 83
    3.1.1 The British CAM Group 83
    3.1.2 The Danish CAD Group 84
    3.1.3 The German CAP Group 87
3.2 Experiences of Collaboration at the International Level ... 88
    3.2.1 The International Data Management Group 89
3.3 Lessons Learned 93
    3.3.1 Constraints on Interdisciplinary Collaboration 93
    3.3.2 Experiences of Improved Collaboration During the Project 96

4 Crossing the Border 99

4.1 Crossing the Border – A Tentative Definition 99
    4.1.1 Fields of Action 99
    4.1.2 Agents Involved in the Shaping Process 101
    4.1.3 Shaping Competence 105
4.2 Participatory Systems Development 109
    4.2.1 Hierarchic–Sequential Design Tradition 109
    4.2.2 The Limitations of Traditional Design 113
    4.2.3 Technology Shaping 114
4.3 The Shaping Process 116
4.4 Prospective Example of the Shaping Process as a Rolling Development 122
    4.4.1 Preparation of the Shaping Process 123
    4.4.2 Future-Creating Workshops 123
    4.4.3 Shaping Workshops 124
    4.4.4 Coordination of the Results of the Shaping Workshops 126
    4.4.5 The Second Shaping Workshop 127
    4.4.6 Seminar with Outside Experts 127
    4.4.7 Conclusion 128
4.5 Summary and Conclusions 128

References 133

Subject Index 136

Introduction

This book discusses and develops concepts and methods of interdisciplinary collaboration between engineers, social scientists and users in the process of developing "human-centred" production technology. The title *Crossing the Border* is intentionally ambiguous as it alludes to three "borders".

The first and most important border which we attempt to cross is the conceptual, philosophical and methodological border which separates social science from engineering design. With few exceptions, the social scientific study of advanced manufacturing technology has concerned itself with the impact of the technology on jobs, skills, productivity, motivation, management style, organization structure, and so forth. In this approach to understanding factory life, technology is treated as a black box and an independent variable. Far fewer social scientists have focused their attention on the engineering design objectives, practices and methods (i.e. the processes) which lay beneath the technology under scrutiny. Similarly, design engineers typically do not concern themselves with the human aspects of their artefacts once completed. Indeed, their training (and the same is generally true for social science training) encourages a bifurcation of the technical and social realms of working life.

In this book we attempt to "cross the border" between social science and engineering design by opening up the black box which so often constitutes advanced manufacturing technology in most social science research and by exploring and developing concepts and methods with which social scientists, engineers and users may work together to design human-centred technology and humane computer-aided work systems.

The second border, closely related to the first, is the border which separates conventional social science research (which takes a reactive stance to the world of technology and work) from applied research and development of a more proactive kind. We take the view that theory must lead to practice in a more direct way than has hitherto been the case in advanced manufacturing technology research. To take a simple example, if empirical investigations show that a given technology directly produces adverse stress reactions in a majority of users, then there is a case for redesigning that technology. The problems which are associated with this shift from reactive observation to proactive design or redesign supply one of the key themes of the book.

The idea of shaping or redesigning technology in a human-centred manner is rooted in urgent social problems reflected in "inhumane" technology and its technocentric application to the labour process. This is not to suggest that the term human-centred is some kind of panacea. The term itself is not unproblematic. For, although it has taken less than a decade for "human-centredness" to become an integral part of the "new factory" and "new production concepts" debate in Europe (particularly within the European Community), the expansive use and definitions of the term threaten to reduce it to a concept synonymous with "user friendly". Hence, one of the key objectives of this book is to clarify such misconceptions through definition (located in the historical critique of production technological development in advanced industrial societies), through the presentation of the fundamental argument supporting the human-centred concept, and by rendering it applicable to practical purposes.

The third border of the title is the most tangible of the three and refers to the national borders (between the UK, Denmark and West Germany) which geographically and culturally separated ourselves as authors. On participating in the ESPRIT (European Strategic Program of Research in Information Technology) project described in later chapters, it became clear that substantial differences of emphasis and philosophy existed between the project participants based in the three countries involved. These differences were manifested not only between ourselves, but also between social scientists more widely and between the engineering designers. For this reason both the project and this book chartered some turbulent waters over their three-year time-span. In the case of this book, we believe that the course which was eventually taken represents a constructive (although by no means definitive or exhaustive) synthesis of social science thinking within the three countries in the area of technology and work. However, as will become apparent during the reading of this book, we do not attempt to hide the differences and conflicts which arose during the project as a result of such differences.

In this regard, the process of writing the book is worth noting. We (the authors) met each other for the first time in 1986 through our involvement in ESPRIT Project 1217 (1199) which began in the same year. As social scientists, we established ourselves as a design subgroup in order to develop a conceptual framework for the concept of human-centredness and to devise ways of translating this framework into usable design criteria or guidelines for our colleagues. Six months later (after numerous meetings in London, Bremen and Copenhagen) we produced a paper on "The Social Shaping of Technology and Work", included in Chapter 2 of this volume.

It soon became apparent that the mere presentation of this paper was an insufficient contribution to the design process. Therefore, it was agreed that each of the three national interdisciplinary design groups should reflect upon and utilize the concepts outlined in the paper. This process was evaluated subsequently and is described in detail in Chapters 2 and 3. However, during the course of this evaluation it became clear that the discrepancy between the concepts within the "Shaping Paper" and actual design practice was far greater than anticipated (or desired).

Thus we were forced to rethink and reappraise the dynamics of the design process (and our role within it). As a result, the Shaping Paper was developed by transforming the concept of human-centredness into a more holistic, participatory and interdisciplinary approach to human-centred computer-integrated manufacturing (CIM) systems. During this process we often had to cease writing and go back to the ESPRIT project and carry out experiments (such as the shaping workshops outlined in Chapter 4) before continuing with our deliberations and the completion of the book. Hence, the development of the human-centred design concept has, to a certain extent, been paralleled by the design methods we devised during the course of writing this book. In this sense the structure of the book mirrors the various stages of our own work process.

Chapter 1 deals with the existing paradigm crisis in the field of technology and work. Three different perspectives are identified and assessed – the technique-oriented, the sociotechnical, and the human-centred approaches – and a work-oriented approach is outlined.

Chapters 2 and 3 focus on the interdisciplinary collaboration which developed within ESPRIT Project 1217 (1199) – entitled "Human-Centred CIM Systems" – between 1986 and 1989. The project personnel were based in three national design groups from the UK, Denmark and West Germany. Chapters 2 and 3 describe the development of these interdisciplinary groups and of the various integrative, cross-national groups which were established. Using a combination of personal experience, project documentation and recorded interviews with project participants, these chapters highlight the differences in cultural tradition which often acted as barriers to collaboration. The importance of developing mechanisms to improve such collaboration is duly emphasized here.

Chapter 4 deals with the problem of design method generalizability and addresses the question of how the experiences gained within the ESPRIT project may be transferred to, and further developed within, other research and development work in the area of computer-aided technology and work. One important issue here is the kind of attitudes and abilities required of social scientists, users and engineers to enable a fruitful collaboration in the design and development of humane computer-aided work systems.

This final chapter also offers a number of meta-rules to guide interdisciplinary collaboration in the design of work technologies. These are exemplified by means of scenarios which demonstrate how such meta-rules may be translated into actual processes of integrated technology shaping. The book ends with a summary of some of our thoughts concerning the essential conditions necessary for fruitful prospective research and development project work.

Chapter 1

The Work-Oriented Shaping Perspective:
The Problem Setting and Perspectives

1.1 Do We Need a New Paradigm?

In his book *The Structure of Scientific Revolutions* Thomas Kuhn (1970) coined the term "paradigm" for conceptual systems that dominate the thinking of scientific communities during certain specific periods of the evolution of science. Initially, each new paradigm has a positive and progressive role. Thus the Newtonian–Cartesian or mechanistic paradigm was so successful in its pragmatic technological applications that it became the ideal prototype for nearly all scientific thinking. Even disciplines like psychology, psychiatry and sociology quite consciously modelled themselves after it.

It is well known that Freud was a member of the Helmholtz Society, whose explicit goal was to introduce into science the principles of Newtonian mechanics. The extreme example is behaviourism – an attempt to eliminate the element of consciousness as a legitimate object of scientific interest. Thus a paradigm may for a period be extremely powerful, clearly defining not only what reality is, but also what it is not and cannot possibly be. However, Kuhn proclaimed, sooner or later continued research will produce data that are incompatible with the leading paradigm. Reality is always much more complicated than any scientific theory, even the most sophisticated one.

At first, all research challenging the dominant paradigm tends to be suppressed. The scientist who generates controversial data is criticized, isolated or even accused of cheating. When the new data hold in subsequent studies and are further confirmed by independent research, the discipline in question moves into a paradigm crisis. Out of the chaos of different alternatives, more or less "fantastic" theories, finally one of these emerges victorious as the new paradigm. This sequence of events is repeated again and again, Kuhn proclaims.

Historical examples of major paradigm shifts are the transition from the geocentric astronomy to the heliocentric system of Copernicus and Galileo and, most recently, from the Newtonian mechanics to quantum-relativistic physics.

As we shall describe in more detail later in this chapter, such a paradigm crisis in the field of work and technology seems to prevail at the moment. At least three perspectives are visible, based on different assumptions of the mutual priority of technical systems and human beings.

Inspired by Numinen, we use the term "technique-oriented approach" to designate the perspective which places the machines or technically mediated communication before the human beings and direct personal communication (Numinen 1988). The term "sociotechnical approach" designates the perspective which seeks a point of equilibrium between the two. Finally, the term "human-centred approach" places human beings and direct personal communication before machines or technical-mediated communication, but still seeks a combination from that kind of priority.

In the following sections of this chapter we intend to describe the above-mentioned approaches in more detail. It should be clear, from the outset, that no single perspective can be understood without reference to the others. At the moment we are unable to formulate a new "victorious" paradigm.

What we can do is to analyse the basic assumptions, strengths and limitations of the three approaches and, from this point of departure, discuss the fruitfulness of a multidisciplinary "shaping" perspective of work and technology. That means a perspective which intends to cross the borders between the technical and social sciences as well as between theoretical and practical knowledge through an action-based dialogue. We describe this concept in more detail at the end of this chapter.

1.2 The Technique-Oriented Approach

1.2.1 Background and Assumptions

The technique-oriented approach is a "child" of the mechanistic model of the universe. This means the imagination of a universe of solid matter made of fundamental building blocks which, by definition, are indestructible. In this universe time is unidimensional, flowing from the past through the present to the future. It resembles a gigantic supermachine governed by linear chains of causes and effects. Therefore it is strictly deterministic. That means that we should be able to predict accurately any situation in the future, if we know all the factors operating in the present.

Similar to the imagination of the universe, the human beings are viewed as biological machines. During the nineteenth century a number of attempts were made to codify and promote the ideas which could lead to the efficient organization and management of workers as if they were machines.

Thus Adam Smith's *The Wealth of Nations* (Smith 1776), praising division of labour, was followed by Eli Whitney's public demonstration of mass production in 1801. Then, in 1832 Charles Babbage, inventor of one of the earliest forms of mathematical computer, advocated a scientific approach to organization (Babbage 1832). It was not until the twentieth century, however, that the mechanical approaches were synthesized in a more systematic way.

Thus the German sociologist Max Weber observed the parallels between the mechanization of industry and the emerging bureaucratization of organizations. Bureaucracy was defined as a form of organization that emphasizes precision, regularity, speed and efficiency achieved through the creation of fixed division of tasks, hierarchical structure and detailed rules for activities. The parallels to machines were clear and deliberate. The mechanistic model proved to be an extremely powerful concept and emerged further during the last four decades, during which information technology has been so widespread in nearly every aspect of working life.

Regarding the field of work and technology, the ultimate dream is the fully automated factory.

1.2.2 Example

The thinking of the technique-oriented approaches may be further illustrated and reflected by an analysis of the ESPRIT-financed research project entitled Design Rules for a CIM System. This study is only one of many, but is chosen as an example here because, at the present time, it seems to be the most systematic approach which clearly demonstrates how human beings as subjects nearly disappear in the technically integrated system (Yeomans et al. 1985).

The principal objective of this study was:

to propose a European CIM system structure. In particular, it was proposed to address three separate but related goals:
1. to modularize the total CIM system into functionally discrete subsystems.
2. to describe the minimum functional specification of each subsystem.
3. to identify the interrelationships that exist between any CIM subsystem and all other subsystems. (Yeomans et al. 1985, p. 4)

In an attempt to do so, Yeomans et al. define three strategies "necessary in reference to any CIM undertaking":

a processing strategy concerning the manner in which processing is to be distributed between a large number of different processing devices;
a data strategy concerning the design and distribution of the total data so that all processors and procedures may have access to consistent and authoritative data values;
a communications strategy concerning the network for a computerized manufacturing system built in such a way that all transmissions can take place within the required time frame. (Yeomans et al. 1985, pp. 7–8)

Though not directly stated, it is obvious that the mechanistic model, or the machine metaphor, dominates Yeoman's way of thinking. Just as an engineer designs a machine by defining a network of interdependent parts and arranging them in a specific sequence and anchored by precisely defined points of resistance or rigidity, so Yeomans et al. intend to "modularize the total CIM system into functionally discrete subsystems . . . and to identify interrelationships between them".

The strength of this approach depends on the possibilities of standardizing and generalizing rules. And this again depends on the stability of the environment and the complexity of the organization in question.

Even Yeomans et al. may doubt the realization of their approach, when they regret that so far "there is no generally accepted list of subsystems". There is not indeed any accepted understanding of the range of company activities which any subsystem should address (Yeomans et al. 1985, p. 2).

As pointed out by Gareth Morgan:

> This approach works well only under conditions where machines work well: (a) when there is a straightforward task to perform, (b) when the environment is stable enough to ensure that the products produced will be appropriate ones, (c) when one wishes to produce exactly the same product time and again, (d) when precision is at a premium, (e) when the human 'machine' parts are compliant and behave as they have been designed to do. (Morgan 1986, p. 34)

If one or more of the above conditions are not fulfilled, the limitations of this approach will be obvious. Moreover, there are several reasons for being sceptical about the "universal mode" which lay behind their approach. In general Yeomans et al. operate with "two quite different types of rules: rules that apply to particular subsystems and interfaces ('design rules') and rules that are generally applicable to all the subsystems of CIM ('Maxims')" (Yeomans et al. 1985, p. 6).

Firstly, it presupposes a universal model of integration of the different parts of the production system (there is always a model behind the "rules"). Thereby an implicit model monopoly and determinism is imposed, characteristic of the mechanistic model as mentioned above. Secondly, generally applicable rules may be an illusion even from a strictly functional perspective of implementing CIM systems. Particular historical contexts of qualification structure in the firm involved, profile of products and existing organization of work, resources available and so on may be of essential importance, when making choices of if or how some parts of a CIM system may be implemented.

Being sceptical about the "maxims" approach is not the same as denying that rules can be more or less generally applied. The essential point is that they cannot be universally applied such as Yeomans et al. intend to do.

1.2.3 "Weaknesses" in the Practical Implementation

The technique-oriented approach to systems integration has so far resulted in an increase in both size and complexity of systems. Efforts have been made to keep systems structure within the limit of controllability. Thus the "top-down principle" (i.e. starting from the system as a whole and then gradually specifying in greater and greater detail until actual implementation is possible) has created new kinds of problems.

The increased complexity of the systems (as they became more and more integrated) has increased the need for documentation too. In order to survey and control this higher degree of complexity, new procedures had to be created which themselves contributed to the increasing number of items to be controlled.

In most cases the technique-oriented approaches have failed to produce the desired results (Majchrzak 1988). More and more it became obvious – even among computer scientists, but essentially among industrial management circles – that reasons for this must lie in the failure to take sufficient account of the reality of work organization and the education of end-users. Owing to a lack of education and perhaps also a lack of motivation, the end-users were continuously found to be insufficiently "disciplined" to handle the systems correctly. Furthermore, old traditional structures of work organization created unanticipated constraints and wastage of resources.

Information technology has grown up in an environment of mathematics and the natural sciences, and has thereby been dominated by the positivistic epistemology of practice. This practice rests on three dichotomies: the separation of means and ends, the separation of research from practice, and the separation of knowing from doing.

Yeomans et al.'s intention to develop universally applicable rules or maxims is clearly an example of separation of means and ends. The whole approach is very abstract and far from the practice of a typical organization. As Numinen (sarcastically) points out: "The model is 'out there' somewhere, awaiting its 'discoverers' like the laws of nature or new continents. The model builder either succeeds in finding the 'correct' model or produces an erroneous one" (Numinen 1988, p. 65).

The weaknesses expressed above do not make the technique-oriented approach obsolete or totally wrong. It means, however, that it is clearly insufficient as the only perspective for implementing new technology in the working life. It is necessary to become more conscious of its limits and allow other perspectives to influence the development.

1.3 The Sociotechnical Approach

1.3.1 Background and Assumptions

The sociotechnical approach can be seen as a response to the challenges, which earlier ones failed to satisfy. The concept was created in the 1950s by members of the Tavistock Institute of Human Relations in England in an attempt to be aware of the interdependent qualities of the social and technical aspects of work. In their view, one element of this dual configuration always has important consequences for the other. This has been well illustrated by many Tavistock studies, for example one conducted by Trist and Bamforth on technological change in coal mining in England in the late 1940s. The assembly-line coal cutting created severe problems – destroying the informal social relations present in the mine. The resolution was to find means of reconciling human needs and technical efficiency (Trist and Bamforth 1948).

The technical and the social systems of an organization are considered more or less equally important; neither system is subordinated to the other. This assumption is clearly a response to the defects of the technique-oriented approach discussed above.

On the other hand, the sociotechnical approach is also a response to earlier work-oriented approaches of more or less mechanistic kind. In the first decades of the twentieth century, Taylor (1911) and other classical management theorists such as Fayol (1949), Mooney and Reiley (1931) and Gulick and Urwick (1937) saw organization of work as a technical problem. Motivation was viewed as a question of the "right" rate of payment.

Much of the work-oriented research strategies since the late 1920s can be evaluated as an attempt to "repair" the social and productive inefficiencies of this perspective. The Hawthorne studies conducted under the leadership of Elton Mayo in the 1920s and 1930s identified the importance of informal and unplanned interactions existing along with the formal organization (Mayo 1933). They showed the importance of paying close attention to the human aspect of production.

Thus the idea of stimulating the productivity through satisfaction of social needs of the employees came in focus for a whole school of organizational psychologists like Frederick Herzberg (1966) and Douglas McGregor (1960). Different modifications of bureaucratic structures began to emerge.

Especially the idea of making employees feel more important by giving them more autonomy, responsibility and "enriched" motivating jobs was articulated as an "alternative" to the technique-oriented and dehumanizing "scientific management". Developed in a number of ways, the idea became well known as a frame of reference for human resource management.

During the 1960s and 1970s, these ideas became attractive to management in an attempt to increase productivity, and reduce absenteeism and turnover, without paying more money. The experiments of Volvo Kalmar in the 1970s and of Volvo Uddevalla in the 1980s present internationally well-known examples of including work satisfaction and health and safety aspects into the increasing technical functionality and quality of products. In parentheses it should be noted that the job redesigns at Volvo Kalmar were highly publicized in the 1970s, but with the benefit of hindsight the results now seem very idealized compared with reality. Whilst it is too early to evaluate the new experiment in Uddevalla, some lessons from Kalmar seem to have been learned.

Niels Björn-Andersen has characterized the basic idea of sociotechnical approach as a threefold one (Björn-Anderson 1980):

1. Job design includes other individual needs of the employees besides pay (hereby clearly distinct from Taylorism).
2. Social and technical needs and goals are considered equally important.
3. Instead of specialized individual tasks, the focus is on autonomous, self-steering groups.

1.3.2 Examples

In the early 1960s the sociotechnical approach was taken up and further developed in Scandinavia (Sandberg 1982; Ehn 1988). The Norwegian Industrial Democracy Project was initiated around 1960, approximately ten years after the initial Tavistock sociotechnical experiments. Between 1964 and 1967 four experiments on work organization were carried out. The solutions were all of the group production type, including some planning activities, but primarily concerned with changes in job distribution and wage systems. The workers' interests gradually turned into resignation, since no important improvement in their situation seemed to come about (Ehn 1988, p. 264).

In the late 1960s in Sweden, and at the beginning of the 1970s in Denmark, experiments were jointly initiated by the central unions and employee organizations. Several interesting ideas of work organization and employee participation came up, but the practical implementation left much behind. For example, the participation for democracy was in no sense included in the aims of the design at the well-known experiment at the Volvo Kalmar plant.

This was one of the reasons why, in the late 1970s, the Swedish union of employees (LO) refused to accept a joint sociotechnical programme for "new factories".

In contrast to the technical-oriented approach, the sociotechnical approach has a much more positive concept of the human being: he is a member of a group. He is generally perceived as an active, responsible person interested in doing "a good day's work", if the circumstances are sufficiently appealing to him. Furthermore, the unofficial communication network is considered just as important as the official one. At least some sociotechnical approaches tried to increase end-user influence and democratic participation.

Thus, the sociotechnical approach contains many aspects which will later be developed in detail in this book. But it also contains several conceptual limitations. Treating the organization as two separated systems in equilibrium of some kind, in which the technical system was largely taken for granted, the questions were largely reduced to more adaptable social relations to the existing technical devices, or the management dominated decisions of the new production technology. The scope for changes therefore was rather limited from the beginning.

In order to be consistent, the ontological status of increasing integration of technical systems and the built-in demand on the social system have never been seriously questioned in the sociotechnical approach.

1.3.3 "Weaknesses" in the Practical Implementation

Partly as a consequence of the conceptual limitations mentioned above, the sociotechnical approaches by and large accepted the technology as given in the practical context too. It improved the working conditions by managerial changes – job enlargement, job enrichment, etc., and particularly (semi-) autonomous work groups. Howard Rosenbrock (a member of the International Committee for the Quality of Working Life in the 1970s) quotes the assessment made by Norwegian sociotechnical specialist Einar Thorsrud: "Many experiments of this kind were conducted and the general experience was that they achieved local success, but did not spread through the organization, and died when the sociotechnical specialists were withdrawn" (personal communication).

The approaches often seem to promise more employee influence and work participation, in theory, than the real power relations in the experimenting plants were able to accept in practice. The workers involved often found that changes stopped before any substantial changes towards real influence had been achieved. Their initial interest or even enthusiasm therefore typically turned into resignation or even hostility towards the "soft" management strategies. As Numinen points out: "Small wonder that these good intentions were often regarded as a form of manipulation, intended to trick the users into accepting the system" (Numinen 1988, p. 104).

On the other hand, this does not mean that the sociotechnical approach should be totally abandoned. As Pelle Ehn noted: "Many of the 'tools' that have been developed are extremely useful in analyzing work organization and production technology: and the job required and group autonomy criteria are, when taken seriously, really a challenge to design for democracy at work" (Ehn 1988, p.270).

The point is that other perspectives may be needed as well. Though the sociotechnical approach in many ways changed the perception of the worker in a more positive direction, it did not really understand the importance of subjectively bounded knowledge in the work process. This aspect is, on the other hand, in focus in the third perspective, which we will analyse in the following section.

1.4 The Human-Centred Approach

1.4.1 Background and Assumptions

In the 1970s the critical evaluation of the above-mentioned limitations of the sociotechnical approaches led to the so-called human-centred approaches. The common core of these approaches was that work culture-based knowledge and actions of human beings should be reflected in a dynamic way instead of being subsumed.

In the UMIST project of a human-centred turning lathe, Howard Rosenbrock expressed his orientation regarding technological development:

> It should not use men and women to perform meaningless fragmented jobs which reduce them to automata. But this is not the same thing as suggesting a return to more primitive craftmanship days. The problem is rather to use the best technology that we know, but to make it an aid to those who work with it, so that their work becomes an enrichment, not an impoverishment of their humanity, and so that the resource which their abilities represent is used to the highest degree. (Rosenbrock 1977)

Rosenbrock attempted in the academic realm to implement the idea of the shapability of technology into a research and development project. He designed an alternative lathe that was not based on a Tayloristic concept. Both the lathe and its CNC (computer numerical control) controller were geared towards preserving and furthering the qualifications of skilled workers.

Rosenbrock's concept of alternative technological development rests on two basic ideas. The first of these he has called the "Lushai Hill effect":

> Our biggest danger lies in what I may perhaps call the Lushai Hill effect. Resting in the evening and looking back over the lower hills, it is easy to say, "At every fork we took the right-hand branch, and see how high we have climbed. Taking the right-hand must be the only way." Though, if we had sometimes taken the left-hand branch, we should in all probability be just as high and perhaps in a region which was in other ways richer, more friendly and more fertile. (Rosenbrock 1980)

The second idea encapsulates the dynamic development of productivity. According to the latter, the immediate reinvestment of rationalization profits into further reduction of the number of skilled workers is bound, in the long run, to prove economically inferior to a strategy designed to put to the company's use the innovatory potential represented by its skilled workers.

1.4.2 Examples

An example in this tradition is the Lucas Workers Plan for Socially Useful Production. As Mike Cooley reflects the process:

What the Lucas workers did was to embark on an exemplary project which would inflame the imagination of others. To do so, they realized that it was necessary to demonstrate in a very practical and direct way the creative power of "ordinary people". . . . The audit of their own skills and abilities, and the surveys in different factories and workshops analyzing and assessing the production equipment, product ranges and skills, represented an enormous extension of consciousness. (Cooley 1987, p. 139)

Lucas Aerospace has had a noticeable impact on the British and international debates about alternative, socially compatible products. Highly qualified engineers, technicians and skilled workers, who lost their jobs following the closing down of an entire production sector, have turned their protest into constructive ideas concerning the production of socially useful items, thereby setting a signal noted in Britain and beyond.

In the UK, this eventually led to a large-scale attempt by the Greater London Council (GLC) to develop a regional economic policy geared towards a use-value-oriented product design. To this end, the London Technology Network was founded. Grouped around the London polytechnics, this aims at combining and putting to community's use the large-scale capacity for innovation represented by the competence of academics, the experience and skills of workers and unemployed people, the needs and desires of the population, of the various trades and of the community services. Technology shaping was viewed as one dimension in the shaping of London's social future, which was to be geared towards the needs, requirements and interests of the many people concerned.

This large-scale experiment rested on a firm belief in the shapability of technology and in the need to govern the latter by means arrived at in a dialogue (and cooperation) involving academics from various disciplines, employees, users, producers and people in need.

Both the London Technology Network (based on an idea of and set up by Mike Cooley) and Howard Rosenbrock's contributions to the shaping of technology represent two important strands from which the international initiative towards a human-centred CIM project evolved. The British ideas and initiatives excited a great deal of interest in Scandinavia and in West Germany among trade unionists and academics, not least because they were grappling with comparable problems.

A Scandinavian project, UTOPIA, at the beginning of the 1980s had a similar objective. A Danish/Swedish research team consisting of both social scientists and computer specialists collaborated closely with the Unions of Printers in Sweden. The objective was the development of a demonstration example of how work quality such as highly skilled labour force, democratic decision process and high level of health and safety are compatible with high product quality. UTOPIA presented an education objective developed in parallel with the alternative systems:

The education should make the printers able to participate in the implementation and adapation of the system. Furthermore it may shape a development process in which knowledge and experiences regarding using the system is transformed to a local further development of techniques and work organization. (UTOPIA Report 1981, pp. 27–28)

In West Germany both IGM (metalworkers' union) and DGB (the top organization of German trade unions) had launched a discussion around the issues of "work and technology", mainly in order to press for a labour and technology policy geared towards a concept of shaping exceeding that underlying attempts at the humanization of work (HdA). True to the tradition of the HdA programme, up until 1983/4, this discussion remained essentially confined to the shaping and organization of the labour process and to the development of those technical items immediately relevant to it.

The first Bremen symposium on Work and Technology provided the platform for academics from the humanities and engineering, for trade unionists, managers and politicians to take stock of the research and development in the area of work and technology and extensively to discuss the concept of technology shaping. Moreover, the symposium afforded the opportunity for a first exchange of ideas with Cooley, who reported on the London Technology Network both in the opening plenary session and in the strand on "production engineering".

The Bremen symposium has stimulated academic debate around the issue of technology shaping and has added the latter concept to the work and technology offensive of the IGM/DGB which is geared towards shaping. This is an important source feeding into the HC-CIM project.

The project initiative was immediately triggered off by a one-week study tour by a German group made up of social scientists, engineers and trade unionists who, following the Bremen symposium on work and technology, discussed with British colleagues the concept underlying the Technology Network, projects concerned with alternative production, and the UMIST research and development project.

The evaluation report bears witness to the ambiguity of the participants' insights, impressions and feelings, having closely analysed the results of that trip. This is borne out by the following extract from the report:

The model of the Greater London Enterprise Board (GLEB) along with its Technology Network is a fascinating experiment, which impresses chiefly by the commitment shown, and the pragmatic approach taken, by the people involved. Leaving aside the economic structures of the UK and EC as well as the pulls and constraints exerted by the world market, the initiators can quote a large number of good reasons in support of their attempt. This consists in a product-oriented initiative tailored to the needs and requirements of the population and the numerous small manufacturers. It is carried out in a close cooperation between academics and people possessing practical knowledge, it is decentralized, aiming at community-oriented innovation around polytechnics. We were fascinated and made doubtful alike by what we saw. Taking into account the economic and political structures of the UK in which these projects are embedded, one begins to have serious doubts as to their long-term viability. Our fascination, on the other hand, was due to the pragmatic commitment and the spontaneity which characterize these projects. This appears to us to be their particular strength and main weakness alike. After evaluating our visit, we were left with a number of open questions:

Can the Technology Network serve as the model of socially compatible technology shaping?

How can alternative economic policy for all London evolve out of individual, cleverly designed products?

How are "alternative" products to be defined in the context of the GLEB model?

Who is the historical subject of the alternative technology and labour policy implemented in the UK, and in London in particular?

Could the alternative lathe developed by UMIST not equally well have been designed in any open-minded company?

At the end of the day, curiosity prevailed over theoretically founded scepticism, and pragmatic research interest as well as common commitment to labour and shaping-oriented research predominated over mere analysis. Thus it was that the HC-CIM project came into being.

1.4.3 Possible Limitations and Challenges of the Human-Centred Approach

Human-centredness is an illusive concept. Behind the concept lies a questioning of the ways in which conventional manufacturing technology has been shaped, and a redefinition of the "socially desirable". In other words, although human-centred systems are technologically possible, tools and methods with which they may be constructed have themselves to be developed.

This confronts engineers and social scientists within the project design team with a dilemma. For although human-centredness can be substantiated and explicated in numerous ways within social science and humanities, the concept does not lend itself to conventional empirical study and analysis. Human-centredness is ultimately a subjective concept which cannot easily be translated into operational criteria.

The tentative focus on a single human being or the man–machine interface which characterized much of the discussion in the UMIST tradition may limit the scope to the micro level. But a human-centred production cell could be incorporated into a work environment based on job simplification and deskilling. As will be described later this dilemma became a quite serious matter of concern in ESPRIT Project 1217.

A challenge to the human-centred approach may be to go deeper into the shaping of the technical systems, thus deviating from the sociotechnical approach's "adaptation attitude". Another challenge may be to integrate theoretical and practical knowledge. But how to do this? As stated earlier there is no "one best way" to do this. At the moment we have no "victorious" paradigm to rely on. So far the human-centred approach contains important perspectives, but by no means an overall frame of reference for the development of work and technology. The "answer" at the moment may be to open the opportunities of new ways of collaboration between different scientific disciplines and people with practical knowledge of the field in question, and thereby gradually develop a combined understanding and strategy for development of work and technology.

1.5 Towards a Work-Oriented Shaping Approach

The "shaping perspective" is fundamentally rooted in the human-centred approach. This means that shaping of technology and work must be in accordance with the values of the human–ecological tradition (enrichment of work quality, use-values of products and ecological balance of energy processes in a broader sense).

On the other hand, the shaping perspective faces the necessity to transcend

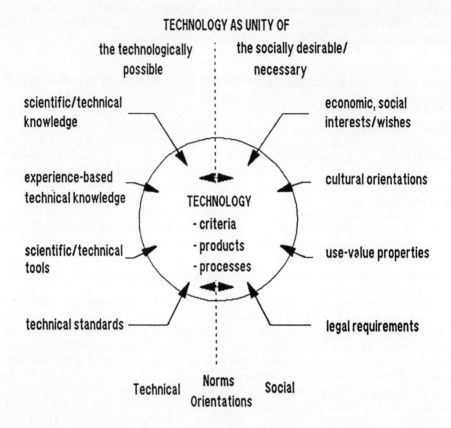

Fig. 1.1. Technology as the unity of the technologically possible and the socially desirable.

the individualistic man–machine perspective which has dominated the human-centred tradition so far, rooted as it is in the skills of the craftsperson.

The challenge of the shaping perspective is not a romantic binding to the past. It is rather a combined transformation of the practical and theoretical knowledge into a new kind of collective skill formation and corresponding software and hardware development, based on the human–ecological tradition.

The accentuation of the "shaping" aspect, rather than "effects", "consequences" and so on, demands a broader concept of technology. Fundamentally, technology is always a union of the technologically possible and the socially desirable. The relation is reciprocal in the sense that the technologically possible and the translation of social purposes into technological artefacts is dependent upon what is technologically possible to construct.

Viewed from this perspective the shaping approach to work and technology becomes a matter of mutual exchanges of perspectives, imaginations, theoretical knowledge and practical experiences in order to re-unite the unfortunately separated "worlds", as a consequence of the powerful domination of the mechanistic paradigm in Western science and technology over the last century.

The idea of reuniting separate worlds may be illustrated in Fig. 1.1. The figure illustrates the necessity to reflect upon the following questions. What are the cultural, social and subjective orientations, wishes and interests governing the design and development of technology? What are the technological possibilities, and in what form are they available in terms of knowledge, tools, methods and abilities? How is it possible to use the different cultural, social and subjective orientations and interests of a multidisciplinary design team in a creative and emancipatory way?

The continuously mutual exchanges of perspectives, so essential for the shaping approach, involve a multidisciplinary approach. Thus we are encouraged to learn how to think of situations from different viewpoints. Every discipline represents a viewpoint often of more or less metaphorical character. For example, the engineering viewpoint will typically be influenced by the machine metaphor stressing precision, speed, clarity, regularity, reliability and efficiency. The psychologist viewpoint may be influenced by the cognitive or psychoanalytic brain metaphor stressing connectivity, redundancy and capacity of self-organizing activities and double-loop learning. The sociologist viewpoint may be influenced by the political metaphor stressing divergent and convergent interests, conflicts and means of power and participation strategies in an attempt to find ways of practice. The anthropologist viewpoint may be influenced by the cultural metaphor stressing the symbolic meaning of day-to-day rituals, language, rules and informal networks and personal relations.

Taken separately, every viewpoint produces one-sidedness. In highlighting certain interpretations or aspects of work and technology, it tends to force others into a background role. Our ability to achieve a comprehensive understanding as a platform for social shaping depends on an ability to see how these different aspects of working life may coexist in a complementary or even paradoxical way.

Thus the technique-oriented, sociotechnical and human-centred approaches may be viewed as reflections of different aspects of work and technology rather than more or less victorious paradigms. Of course, this is not to deny that they also have the functions of paradigms. What is being stressed here is that the dogmatic use of only one of these perspectives may result in narrow-minded or idealistic research.

The human-centred approach has rediscovered the importance of subjectively bounded knowledge developed through mutual face-to-face communication and practical learning-by-doing. The position of the individual is clearly more prominent than in the technique-oriented and sociotechnical approaches. The meaning of work is sought within individuals themselves. If a person is allowed to control and influence his or her way of working and its results, he or she will also show responsibility and produce products of sufficient quality. Responsible labour is a natural human activity. Therefore technology must be shaped in a way which supports the "whole person" to use and develop creativity, pleasure of work and special abilities.

The sociotechnical approach has rediscovered the productivity-enhancing effects of semi-autonomous working groups and informal networks of commuication. Moreover, the importance of user participation as a means of gaining acceptance, improving the quality of the implemented technology and tentatively increasing the democratic effort to enhance the influence of groups at the shop-floor level has also been stressed, though not always successfully practised, by this approach.

The technique-oriented approach has been the "Prugelknabe" (or "whipping boy"), because of its deterministic world view and obviously reductionistic understanding of the human being.

On the other hand, this approach has introduced a number of important questions too. It has focused on aspects such as the problems of overall integration of different work areas and the rigidity of the traditional division of labour, and has generally broadened the scope by developing more or less sophisticated models of analysis, in particular the functional relations of planning and production as a connected system. While the human-centred approach traditionally has focused on the individual level and the sociotechnical approach has focused on the group level, the technique-oriented approach has more and more developed models for overall system integration.

Though these models may be criticized from different viewpoints, they have also provoked social scientists to realize limitations and new possibilities in their own understanding of work and technology. Therefore, these models are also important ingredients of a continuous shaping process.

Rather than discussing the superiority of one of these approaches at the general level, the shaping perspective prefers to create a mutually stimulating and fruitful perspective and development environment through an open multidisciplinary, action-based research teamwork. As Morgan (1986) points out: "Rather than impose a viewpoint on a situation, one should allow the situation to reveal how it can be understood from other vantage points" (p. 337).

In order to do this, new methods of dialogue are needed. The stressing of dialogue is not accidental. The shaping perspective differs sharply from the "objectivistic" paradigm dominating technical-oriented approaches. Individuals are perceived as objects of modelling like anything else. If anything, they possess fruitful knowledge and potential resources of transcending limitations of work. They are subjects that are able to act and participate in a dialogue. Experience is analysed as if one's own experience were an object outside of oneself. Rather the "outer" and "inner" world are mutually fabricated in an active reciprocating process. The researcher is part of the shaping process, not standing outside as a "neutral observer". Therefore research is not only a question of analysing and evaluating, but also shaping new visions, processes and products in a continuous dialogue with people involved. Thus the shaping perspective necessitates crossing different kinds of "borders" established both in the scientific world and between theoretical and practical knowledge.

The multidisciplinary approach may be an obvious opportunity to practise such an approach. However, it presupposes an open-ended mode of enquiry which enables problem setting and possible alternatives to emerge from the dialectical shifting of viewpoints or metaphors, rather than a mode of enquiry which focuses on isolated problems and/or piecemeal solutions.

The real challenge, then, is to develop social and technical conditions and qualifications for such an open-ended mode of enquiry. This challenge includes the practical experience of work. User participation sounds obvious, but may also involve a potential risk. As the Norwegian sociologist Stein Braten points out:

If a model strong actor and a model weak actor are coupled in an open information exchange system, the former may be expected to gradually increase his control of the other actor. Offers of information are useful only to the extent that there is model capacity for processing the information offered. Thus a successful transition in the name of democratization to a more open communication structure may freeze – or even increase – the influence gap. (Braten 1983, p. 190)

User participation in a work-oriented shaping process sounds obvious. But how to do it in practice? This is not only a question of more sophisticated methods and techniques, but also of common understanding and trust. The means ought not to be separated from the needs, as many experiences from different sociotechnical approaches from the 1960s and 1970s may have taught us. Or more precisely: user participation may include ends as well as means to reach the ends. If user participation has only an instrumental character, the feeling of manipulation and frustration will prevail sooner or later. Therefore the real core of the problem is how to establish the context and concrete methods by which user participation, in its most comprehensive form, can take place? This aspect will be further discussed in the following chapters.

1.6 Summary

As a summary the following questions give some insight into the common set of orientations and values behind the work-oriented shaping perpective.

How to respect and reflect work culture-based knowledge and actions of employees and at the same time transcend rigid status traditions (e.g. white- and blue-collar workers) in order to achieve new kinds of collective skill formation?

How to encourage a "bottom-up" participative approach and at the same time impose a holistic vision of better integrated overall process of organization and technical means of function?

How to increase the capacity for self-organization and innovations emphasizing the special human abilities of learning from errors or deviations and question the rationality of the existing norms or rules of behaviour and at the same time include the technical opportunities to shape new ways of productive and qualitative work procedures?

How to take into consideration that a changing process is not only change of technologies, structures and abilities of employees, but also images and orientations of management?

How to develop multidisciplinary approaches to the shaping of technology and work that reintegrate the routines and imagination fantasy of skilled workers and designers and at the same time use the technological possibilities as much as possible to support the efficiency of this process?

Though we do not believe we have the "final" answers to the questions, the following chapters in this book may stimulate further clarification and understanding of how to deal with the work-oriented shaping approach both in theory and in practice.

Chapter 2

The Interdisciplinary Design of Computer-Integrated Manufacturing Systems: ESPRIT Project 1217 (1199) Part One

2.1 Introduction

The following two chapters focus on the interdisciplinary design processes which developed in ESPRIT Project 1217 (1199).

The project, which began in May 1986, was based on the premise that a computer-integrated manufacturing (CIM) system will be more efficient, more economical, more robust and more flexible if a person is directly in charge than a comparable unmanned system. The project objectives were fivefold:

1. Establish criteria for the design of human-centred CIM systems.
2. Establish their economic and commercial competitiveness.
3. Achieve a high level of flexibility and robustness in CIM systems.
4. Define the training for a new type of multi-skilled worker.
5. Demonstrate at a number of production sites that there are better means of organizing manufacturing, especially suited to Europe.

Because the general orientation of the ESPRIT CIM initiative was towards the ideal of the peopleless factory, the acceptance of the project hinged on the committed support of trade unions (particularly the metalworkers union in West Germany) and on the competent presentation of a research proposal by engineers. However, the basic idea of human-centred technology shaping quickly gained support within the ESPRIT programme as it reflected a growing concern within the Commission of European Communities that Europe should develop its own manufacturing technology (rather than rely on imported technology from Japan and the USA).

The initial proposal, submitted in 1984, entitled "Comprehensive criteria and exemplary prototype of a human-centred CIM system", was rejected by the ESPRIT directorate. The subsequent, less ambitious, proposal was accepted although funding was forthcoming for three rather than the proposed five years. As a result of this more limited funding, it was decided that the design and implementation of a full-blown CIM system at a user site was not possible. Instead, the partners would be grouped into three separate, but interlinking, national groups: a computer-aided manufacture (CAM) group in the UK, a computer-aided design (CAD) group in Denmark and a computer-aided planning (CAP) group in West Germany. Each group would be responsible for building and demonstrating their particular CIM module.

Each of the national groups contained social scientists as well as engineers, some with a university background and others with a software development or industrial company background. Coordination of work was the responsibility of a board comprising representatives from each partner.

As well as being situated in different countries, the three groups had different cultural and philosophical perspectives and traditions which they brought to bear on their design work. Moreover, during the project both an International Social Science Group (ISSG) and a more technical International Data Management Group (IDMG) were established. These two groups contained representatives from each of the three countries.

With detailed reference to the ESPRIT project, the following two chapters will describe and analyse the following themes: how the "modus vivendi" and especially the interdisciplinary collaboration and conflicts developed in each of the national groups; how these trends may be interpreted in relation to the historical context of the three groups, particularly in the first year of the project, and in relation to the influence of internal group dynamics and the ISSG and IDMG as the project progressed; and what lessons may be learned from these experiences of interdisciplinary collaboration with regard to the development of human-centred technology.

The ensuing discussion follows the events which occurred during the three-year life-span of the project in chronological sequence. Although this tends to give the discussions something of an open-ended feel, it has the advantage of giving the reader a clearer idea of how the interdisciplinary design process developed over time. In this chapter the experiences of interdisciplinary design during the first part of the project are outlined. This includes a full version of the original "Shaping Paper" written by the authors. This paper was to prove highly influential in focusing ideas and design methods in all national groups and is included here accordingly (see section 2.4.2). As will become apparent, our own ideas were to develop beyond those expressed in the "Shaping Paper".

Sources for these two chapters include official project reports, discussion documents, published papers, minutes of meetings, project correspondence, tape-recorded interviews with key participants, and notes taken at workshops by the authors and others.

2.2 Experiences of Interdisciplinary Collaboration During the First Year of the Project

2.2.1 The UK CAM Group

At the beginning of the ESPRIT project the CAM group comprised three partners: the Greater London Enterprise Board, UMIST, and the engineering subcontractor company RD Projects Ltd. The remit of this group was to design, develop and implement a human-centred CAM turning cell as part of an overall CIM system. The cell was to contain two computer numerically controlled (CNC) lathes, a cell controller and automatic parts and materials transfer equipment.

At the start of the ESPRIT project, the CAM group generally felt that the knowledge and expertise which had been gained by the UMIST researchers could be readily transferred to the CAM group work. In addition, the RD Projects engineers had considerable experience in designing machine tool programming hardware and software. It therefore seemed that the two areas of expertise could be combined to form a solid foundation upon which to develop a multidisciplinary design method. This optimism was reflected in the absence of any provision for the employment of a full-time social scientist in the CAM work.

The first six months of the project witnessed close collaboration between engineers from UMIST and the subcontractor in an effort to combine the best features and functions of the UMIST lathe programming and cutting technology software with the best features of the programming software and hardware developed by the subcontractors for one of their own machine tool controller designs.

This collaboration took the form of monthly design meetings at which a social scientist was usually present in an advisory capacity. This group felt that the most fruitful input of human/organizational considerations could be made once a prototype had been developed during the first year. During this time no user company was formally contracted in the project.

Within six months of starting the CAM work it became clear that the design process was becoming increasingly dominated by technical considerations. For example, the choice of the two CNC machines and the work-handling equipment for the CAM turning cell was made without consultation with social scientists. Part of the reason for this technically led approach to design stemmed from the subcontractor's engineers' disregard for the social science input because it was "lacking in specificity". Conversely, from the social science viewpoint, the chief obstacle in the CAM work was the engineers' view that, unaided, they could produce a human-centred system.

The relationship between the social scientists and engineers became increasingly acrimonious, thereby creating a project management crisis. The crisis was resolved through the personal intervention of Rosenbrock, who initiated a series of weekly CAM group design meetings during which the engineers presented details of the major design options available at each stage of design (e.g. choice of work handler, software environments, and human-machine interface design). The options available at each stage were then examined in detail by the group before a final decision was agreed.

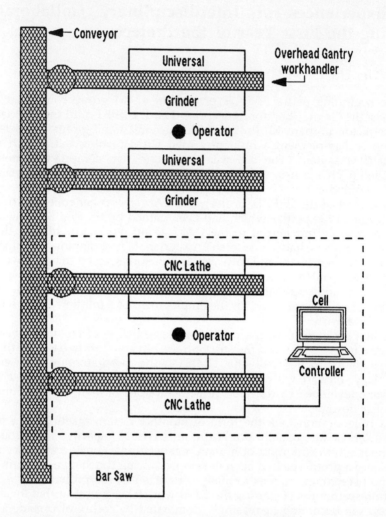

Fig. 2.1. Proposed layout of the computer-aided manufacturing cell.

Although extremely time-consuming, this series of meetings between December 1986 and April 1987 resulted in the establishment of a relatively harmonious relationship between group members and a mutual understanding of the methods and constraints associated with different perspectives. As a result many important design choices were unanimously agreed by the group. The overall technical layout of the CAM cell, for example, was agreed and is shown in Fig. 2.1. Note that the provision of one operator to work on both CNC lathes was to remain a focus for debate between the social scientists and engineers (see below).

However, there were notable disagreements within the group over the nature of the human–machine interface. These disagreements polarized between the engineers' preference for manual data input (MDI) by push-button menu selection and the UMIST team's preference for MDI by analogue

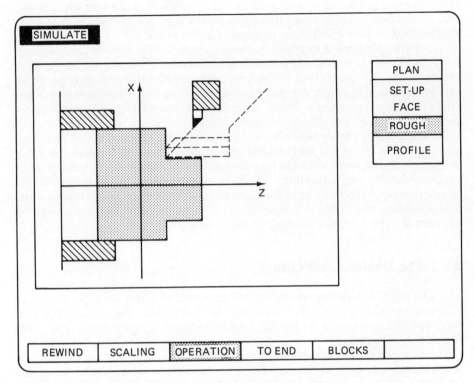

Fig. 2.2. Visual display screen layout for the CAM lathe controller.

pointing device (e.g. tracker-ball or electronic hand-wheels). The issue was resolved through the adoption of a WIMP (Windows, Icons, Mouse and Pull-down menus) interface using push-button menu selection rather than an analogue device (a mouse). An example of the resultant human–computer interface screen layout is shown in Fig. 2.2.

This compromise solution was accepted by the UMIST team owing to the engineers' assertion that no analogue device was currently available which could both withstand the rigours of a shop-floor environment and was accurate enough to enable detailed part program editing. This compromise solution was accepted although all parties agreed that it was far from ideal.

On the UMIST side it was felt that the opportunity to explore the possibilities of record–playback programming (Noble 1984) had been lost, as was the opportunity for operators to more directly manipulate machine tool control systems. The engineers, on the other hand, were aware of the time-consuming nature of such exploration. In addition, more detailed discussions revealed that the engineers were also evaluating design options (such asthe use of analogue devices) in terms of their own implicit criteria. Of these criteria, the criterion of "consistency" was the most referred to in connection with design decision-making.

The engineers' interpretation of consistency focused on the need for interface controls and input devices to be used for as many related functions as possible. For example, the criteria are infringed in designs where an alphanumeric keyboard is used for data entry, a mouse is used for text editing, and an electronic hand-wheel or a light-pen is used for geometric data entry and editing. A better solution, in this example, is the use of just one device to carry out all data entry and editing (when technically or economically feasible). When it is pointed out that such a criterion could not be justified on ergonomic grounds, two of the engineers replied that this was probably due to the absence of any ergonomic studies in the area. It was clear that the criterion of "consistency" had a powerful subjective, even aesthetic, dimension to it in the minds of the engineers.

As the project progressed the collaboration between the social scientists and engineers improved and became more productive. As a result the CAM design meetings were held far less frequently to enable development work to progress "more smoothly". It was agreed that in addition to the development of decision aids for the calculation of cutting parameters and tool selection, the CAM cell operator(s) should be provided with software support for the microscheduling of work within the cell.

2.2.2 The Danish CAD Group

The Danish CAD Group comprised three partners: the Institute of Social Sciences at the Technical University of Denmark, the Department of Mechanical Engineering in the Technical Institute, Copenhagen, and NEH Consulting Engineers.

The remit of the group was to optimize "the alternative design methods to bring about the best from the designer, accommodating each designer's preferred method of working" (Technical Annex to ESPRIT proposal 1985).

During the first six months of the project, the social science and engineering groups within the overall CAD group worked separately with very few established lines of communication and discussion. The engineers began work by evaluating state-of-the-art flat panel displays, graphic drivers and digitizers. The Hanover Fair was visited to closely examine products similar to the electronic drawing board concept which lay at the heart of the CAD work. Quite independently, the social scientists established three user groups comprising technical school teachers, industrial designers and industrial draughtsmen.

At the first meeting between the social science and user groups the project's objectives were introduced and mutual expectations and benefits were discussed. At subsequent meetings, the positive and negative attributes of conventional CAD systems with regard to the design process were debated, in addition to possible principles for developing prospective job structures and work organization.

During the first six months of the project 12 all-day meetings were conducted with these three user groups. It is worth noting that the project engineers did not participate in these meetings. They expected that the social scientist group would transform the results of the user group discussions into an operational form for subsequent software development.

The meetings produced a detailed number of criticisms of existing CAD systems together with a list of suggestions for their improvement. This list was presented to the technical groups at an early stage of the project, although little notice was taken of these suggestions as the engineers perceived their primary task at this time to be the development of a drawing board prototype and CAD system interface hardware/software.

The first meetings between the social science and engineering groups were held in an atmosphere of distrust, scepticism and misunderstanding. Hence the social science group feared that the engineers would develop their technical specifications and models too quickly, thereby making important technical decisions before the concept of human-centredness, and its related criteria, were developed.

On the other hand, the engineers feared that the social scientists and the user groups would only generate unrealistic demands concerning technical specifications and software development. One engineer viewed human-centred CAD merely as something different to systems currently on the market. Another engineer defined the concept as "a question of more or less user-friendly". The general feeling appeared to be that once human-centredness was confronted by criteria of technical feasibility it would "not be much different from the conventional automation approach".

Moreover, the engineers feared that the social scientists would play the role of controlling their work without clearly specifying the criteria of human-centredness. Therefore they continuously asked the social scientists to precisely define these criteria. During the early stages of the project the social scientists were not able to do this, thereby increasing the prejudice of the engineers towards a belief that "human-centredness is scarcely anything but a smart way of getting money".

After six months of independent endeavour the engineers presented their ideas at a meeting to which all user group participants were invited. The technical groups had considered six different possible designs and presented the "best solution" to the interdisciplinary group meeting. This consisted of a more advanced drawing board design (see Fig. 2.3). This drawing board design includes a drawing machine, a display and a digitizer. The display is of the flat panel display type showing the menu as an item and a text field, together with the geometrical contour of the drawing. The pencil is wireless so as not to disturb the user's free hand movements. A digitizing eraser makes it possible to erase any part of the drawing. The drawing board was intended for use in three modes – free-hand drawing, digitizing and keying input – from which the designer was free to choose his or her preferred method of working. Whilst drawing, the display shows the window of the drawing at the tip of the stylus (on a 1:1 scale if desired). The idea is to reduce the "keyhole" effect associated with some conventional CAD systems.

The users' response to the design was not very positive, partly owing to misunderstandings of the engineers' intentions, and partly because some of the users did not approve of combining CAD and drawing boards in the same system. It was therefore decided to arrange a user group meeting with only one representative from each of the three user groups present in an effort to go more deeply into the problem. At this meeting the representative from the designers' user group put forward an idea that was to have a significant impact on the CAD group. This designer suggested that the CAD group should develop an electronic sketch pad. This would be portable to allow the designer to discuss aspects of a product's design with people elsewhere in

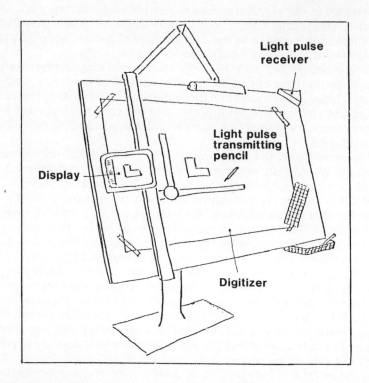

Fig. 2.3. Advanced prototype computer-aided design system.

the organization. Such a sketch pad would greatly facilitate this process of interpersonal communication and discussion. Furthermore, the designers' user group representative seemed interested in testing out the idea in a practical organizational setting.

This suggestion was an important catalyst to the work of the CAD group for two reasons. First, it offered a specific user-oriented tool as a clear goal for the engineers to work towards, thus motivating them to develop prototype software and to find appropriate hardware to achieve this goal. Second, it improved the relationship and collaboration between the social scientists and engineers. The presentation of a clear product triggered off intensive discussion between the different disciplines concerning the technical/economic feasibility of the sketch pad idea, its role as a medium of face-to-face communication and how the research and development work should progress and be organized.

The design "shift" occurred in September 1987 during an all-day meeting between the leaders of the three CAD groups, where it was decided that the electronic sketch pad suggestion put forward by a representative from the industrial designer user group represented the ideal design solution for the CAD project groups. After this meeting, the collaboration between the Danish groups began to develop in a positive and more constructive manner.

2.2.3 The German CAP Group

The German CAP Group was based at the University of Bremen and comprised three research and development groups located in the departments of Vocational Education, Production Systems Design and Organization, and Production Technology. In addition, the industrial company Krupp Atlas Electronik participated as an industrial partner.

Before the start of the project, it was planned to allocate the research associates to the respective heads of these departments whilst taking them on as a group to encourage interdisciplinary teamwork. Unfortunately, it transpired that the academic interests of individual professors prevailed, so that vacancies were filled by specialist researchers (engineers, ergonomists and vocational educationalists).

During the first six months of the project, the CAP group encountered only minor difficulties in collaboration. One unanticipated difficulty which faced the group during the early months of the project was caused by the different work practices and jargon used by the different specialisms. While the engineers consistently presented, communicated and documented their ideas through the use of diagrams, the social scientists and educationalists were more used to presenting their ideas in the form of written text.

The expectations of the engineers with regard to the work of the social scientists was summed up by one professor as follows: "I suggest that, independent of our common discussion, we [the engineers] develop a practical proposal for a CAP solution. In the next step, this should be critically analysed by the social scientists from all possible angles. This is their strength anyway."

The work of the CAP was based around the concept of competence-oriented workshop control, production islands, and qualification. The group was largely in accord with these ideas. Indeed, the CAP group was to spend a significant amount of time engaged in productive discussion concerning human-centred CAP philosophy.

In the course of these initial discussions, the rough structure of the international division of labour in the ESPRIT project between the CAP, CAD and CAM components was modified and a general direction of development was devised. The initial view of the interrelationship between the three components as a simple hierarchy (see Fig. 2.4) was altered into one symbolizing the transfer of managerial and operative planning competence to the shop-floor level (see Fig. 2.5).

Turning Fig. 2.5 through 90 degrees is to express that the hierarchical, centralized factory model is to be replaced by one in which production and administration (logistics), organized in islands, cooperate with each other according to the releases of superordinate production planning (see Fig. 2.6).

In designing such decentralized computer-aided production planning systems, the CAP group was faced with the question of how much planning should be delegated to the shop-floor. Mertens (1985, p. 102) distinguishes between three levels of planning:

1. Production programme planning – long-term planning which defines company targets.
2. Manufacturing programme planning – medium-term planning which defines strategies for achieving the targets set by the production programme.

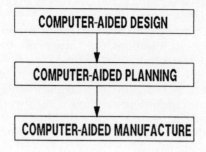

Fig. 2.4. Initial computer-integrated manufacturing hierarchy.

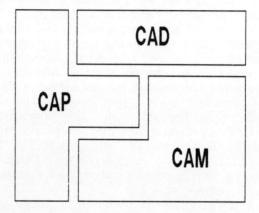

Fig. 2.5. First modification by the computer-aided planning group.

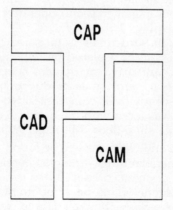

Fig. 2.6. Second modification by the computer-aided planning group.

3. Workshop control – short-term planning of how targets will be achieved on the shop-floor.

It was envisaged that both workshop control and elements of manufacturing programme planning would be based at the shop-floor level in the ESPRIT system through the adoption of the "production island" approach to production systems design. This concept of the production island was developed by Brödner (1985) to describe the organization of small-batch production according to the principles of group technology in which families of products or parts are manufactured in their entirety by semi-autonomous work groups of highly skilled people. A more detailed outline of this concept may be found in section 2.4.2.

The CAP group considered that this concept opened up the path to the future of production work and technology which is most conducive to turning CAP, CAD and CAM into human-centred tools. In addition, there were technical and economic arguments which supported the island concept:

1. Orders being processed within a production island often do not require such detailed process planning as in traditional manufacturing environments. It is sufficient to specify the due completion date for the component and the drawings, and not the sequence of all operations.
2. The order lead-time is essentially shorter, and last-minute changes of delivery dates as well as urgent jobs can be handled with fewer problems.
3. As a result of the higher qualifications of the island workers, some design changes can be handled more easily, since the worker can often work from a rough draft.

These arguments (especially point 3) indicate that the abilities of staff must be taken into account as an independent variable in shaping the CAP concept.

Therefore, the CAP group were in accord over the theoretical underpinning of their work. However, the group confronted problems with the issue of user participation which had been raised by the work of the CAD group in the early part of the project. With the exception of that psychological position in work science which is fixated on the analysis of work processes, the majority of the CAP group members were quite open to the idea of participation as an important dimension in the shaping process. Nevertheless, there were objective and subjective barriers to its realization.

First, setting up a user group for the development of CAP tools would have required the recruitment of technical employees mainly from operations planning because, in factories, this is where the planning and controlling of the production process typically take place. Hence the involvement of planning personnel would have clashed head on with the CAP group's objective of reintegrating important functions of operations planning on the shop-floor. It was seen that all the technical employees who have sufficient experience with planning and control are to be found in operations planning and were therefore not eligible to become members of a user group.

The CAP group work aimed to develop a different model of work organization. This, however, had to be achieved without involving those immediately concerned because they all lacked the necessary technical knowledge and experience.

A second barrier to user participation was the CAP group's lack of experience with user participation, although group members did not see the lack of a user subgroup as a major problem. The contradiction between the comparative openness to (if not interest in) user participation on the one hand, and the lack of a corresponding group initiative on the other, cannot be attributed to the objective barrier outlined above. The factory model, based on the principles of island production, as well as the concomitant conditions for computer-aided planning, were, from the outset, unanimously endorsed by the entire CAP group and understood as being in the objective interest of prospective users. According to this logic, the integration of peoples' subjective interests is less important. All the CAP group members concurred on this point.

The design work of the CAP group during the first half of the project was additionally troubled by the nature of the relationship between the academic and industrial partners. In the day-to-day work of the group, their collaboration was hampered by a number of factors. First, the industrial company engineers were not accustomed to elaborating concepts such as human-centredness. Also the CAP product had no specific customer in mind. This called into question the validity of engaging in a project where the customary criteria and procedures of project acceptance were lacking. As a result, the industrial partner did not fully participate in the CAP group discussions until something resembling a product was developed.

This lack of cooperation during the first half of the project resulted in something of a project dualism within the CAP group. Whilst the academic partners were elaborating and concretizing the concept of human-centredness, the industrial partner began developing an existing shop-floor monitoring and control (SMC) system according to the principles of conventional industrial design. As a result, two different workstations were developed within the CAP group. The industrial partner opted for a more conventional design solution whilst the academic partners opted for DISCUSS, a production island-oriented shop-floor control system.

This dualism did not involve a conscious decision against collaboration. Under time pressure, the academic designers (like their industrial counterparts in Krupp Atlas) resorted to existing solutions. The involvement of social scientists in the actual design process served to increase the designers' dilemma of wanting to create something new whilst having to start from something which was already available. The internalized principle that the "system has to be finished" thus heightens the barrier which stands between social scientists and engineers collaborating in the actual design process. Yet despite the fact that both the academic and the industrial engineers felt bound by the same objectives and developed SMC designs based on pre-existing designs, two different workstations were eventually realized. One of the reasons for this is the fact that industrial designers have to gear their efforts to marketable products and that their scope for acting is circumscribed by design routine. Academic designers, by contrast, have comparatively more scope for selecting suitable designs.

One final aspect of the CAP work which deserves mention relates to the constraints on interdisciplinary design imposed by the hierarchical nature of German universities. A number of problems were created by the prominent position of the professors vis-à-vis the research associates and by the dependence of non-professional academics as laid down in the University Act and in researchers' work contracts.

The rule that research associates must work subject to instruction, if consistently applied, tends to reduce initiative and independence among subordinate group members. If, by contrast, this rule is informally ignored (i.e. research associates are allowed to carry out research with some degree of independence) this may result in role conflicts that may equally impede the research and design process. In the latter case, there opens up a wide gap between the formal responsibility and the actual competence of the professors. They may have to assume formal responsibility for research work with which they are insufficiently familiar and have contributed little towards. Both trends had some impact on the output of the CAP group during the first half of the project.

2.3 Evaluation of the First-Year Experiences in the Three National Groups

2.3.1 National Differences

As described above, the three national approaches had quite different points of departure owing to differences of historical context and background. For example, the German CAP group were heavily influenced by the debate on new production concepts which began in Germany around 1980. In this debate, some fields of production engineering as well as researchers in engineering science and industrial management have increasingly turned away from rationalization strategies that are exclusively geared towards scientific management (Taylor 1911).

Structural changes in production posed a new challenge to production planning systems as improved strategies are needed to enhance a company's market share, service level and flexibility. Industrial management were turning their attention increasingly towards controlling the flow of work as a means towards productivity and profitability. The new production concept debate in Germany and elsewhere (Brödner 1986) questioned the validity of centralized production planning and instead focused on qualification and the decentralization of production responsibility to the shop-floor.

The CAM group, on the other hand, were heavily influenced by the UMIST Project (1982–85) under the direction of Professor Rosenbrock (Rosenbrock 1989). The original aim of the UMIST project was to develop a complete flexible manufacturing system (FMS). Unfortunately, because of lack of supportive funding, the hardware for this proposed system was not developed. Accordingly, the project focused on the design and development of a CAM turning cell rather than a full FMS.

By the end of the UMIST project, sophisticated software had been written and coded to enable skilled turners to write and edit complex part programs for a lathe. Many features of this software, including a shop-floor decision support system, were incorporated into the CNC lathe controllers developed in the ESPRIT project.

Thus, the CAP group primarily focused on the work organization level of analysis, whereas the CAM group focused on the human–machine interaction level. The Danish group shared a similar human–machine level perspective with their British colleagues, although the Scandanavian research tradition of user involvement and participation in design was more in evidence in Denmark than in the UK.

The Danish social scientists were theoretically inspired by Habermas' concept of "communicative action" and Polanyi's concept of "tacit knowledge". In addition, the work of Cooley and Rosenbrock was to influence the group's way of thinking. Coupled with their research on the impact of technological change on industrial designers and draughtsmen, this outlook stressed the importance of relecting on practical skills and the more or less conscious relation between the mental and physical processes in evidence during design work.

The stress was therefore upon ensuring that tacit knowing is still needed and is more highly prioritized as CAD technology develops and on improving the physical and social context of the designer to enable a reliance on his or her own qualified, subjective judgement. The core of the CAD group's research was therefore the designer's possibilities to preserve or widen his or her subjectively based initiative during the designing process.

It is worth noting that, although practical experiences and subjectively bounded skills played a central role in the general frame of reference of all three groups, this was actually practised at the methodological level only by the CAD group. Neither the CAP or CAM group involved end-users directly in the exploratory, developmental process at the beginning of the ESPRIT project. On the other hand, the German work organzation perspective was important for the further development of the project, as we shall see later in this chapter.

2.3.2 Critical Evaluation of Interdisciplinary Collaboration During the First Year of the Project

It was commonly accepted during the ongoing interdisciplinary dialogue that any successful techological development, and particularly one claiming to be human centred, requires general orientations regarding humaneness and social compatibility. However, neither social scientists nor engineers know how these general orientations can be made to bear fruit in the development process. The problem of mediating between basic ideas and general orientations on the one hand, and tangible shaping criteria and dimensions on the other, renders any interdisciplinary dialogue difficult because the engineers will always insist on the tangible level, while the social scientists will always resort to the level of criteria which may be difficult to operationalize. One important task to be fulfilled by interdisciplinary research and development therefore consists of a mediation between these levels (and world views) and of making participants aware of this mediation, thereby turning it into a continuous process.

Given this, and with the benefit of hindsight, the separation of engineers, social scientists and users into distinct groups within the national project groups was a mistake. For example, within the CAD group it is now realized that it was a mistake not to involve the technical groups in the user group discussions from the beginning of the project. The initial reason for their exclusion was to avoid the premature closure of design perspective and options open to the users. It was feared that the engineers' special interest in creating an "easy to use" CAD system would exert a strong influence on the users. But the unanticipated side effect of this disaggregation between the user/social science group and the technical group was an increasing "gap" in perspectives.

This gap became evident towards the end of the first year of the project when the technical group presented their idea for a combined CAD system and drawing board. This idea met with little enthusiasm from the user groups (see section 2.2.2). On the other hand, the technical group realized that they had committed themselves to the idea of the electronic drawing board virtually from the onset of the project without any consultation with users.

The CAM group experienced a similar gap in perspectives, although this was never reflected by the participants. When conflicts arose within the CAM group (e.g. the analogue versus digital data input discussed in section 2.2.1), it was commonly viewed as a conflict between individual aesthetics and predispositions rather than a more deep-rooted conflict between design philosophies. It is unlikely that shaping criteria, no matter how specific and detailed, would have resolved this conflict (Corbett 1987).

Within the CAP group collaboration was essentially moulded by three research traditions: engineering–scientific pragmatism, psychological objectivism, and pedagogical reductionism. The "gap" between these traditions stemmed from the comparative lack of theory and specific methodologies to guide prospective design problem-solving in engineering pragmatism, and the inability of the social scientists to contribute to the design work without first having a finished product to analyse.

Thus, the educational scientists were primarily concerned with the shaping of training and education processes. This orientation facilitates collaboration with other disciplines, yet the pronounced didactic approach to the explication and solution of problems produces an implicitly determinist understanding of human-centred CIM in which technology and work are taken as given. Instead, qualification requirements and educational contents are derived on the basis of technological products. Because of this reactive approach to CIM system design, the pedagogical scientists made little impact on the CAP design process.

Similarly, the psychologists within the group tended to share a mechanistic view of work closely related to the position held by their engineering colleagues. Work psychology is based on fairly comprehensive theories and rules (e.g. Hacker 1986), which makes interdisciplinary collaboration all the more difficult if other disciplines cannot be subsumed into such theories. As with the pedagogical sciences, this bias towards the analysis of work and the assessment of technological givens acted as a barrier to inter-disciplinary collaboration and to prospective technology shaping.

Prospective interdisciplinary design shaping therefore should place much more emphasis and effort towards joint workshops from the beginning of the design process in an attempt to bridge the "gap" between the different visions and perspectives of the project partner disciplines. In the same vein, much more time needs to be given at the start of the collaboration process to explicate and clarify these differences. It is from this initial process that appropriate shaping methods must develop.

2.4 The "Shaping Paper" – Intentions, Reactions and Influence

2.4.1 Intentions: The International Social Science Group (ISSG)

The ISSG was officially established in August 1986 and comprised one social scientist from each of the three national groups. The objective of this group was to formulate a common frame of reference and suitable methods regarding human-centred perspectives on technological development. Although sharing a similar social science perspective, the three members (the authors of this book) came from different cultural traditions which, particularly at the onset of the project, created a number of hard and intense discussions before a common understanding was reached between them.

Six months later (February 1987), the ISSG presented the Shaping Paper to all partners in the project. The ISSG viewed the paper as a point of departure for a dialogue with engineers in the project. Thus, it was hoped that the Shaping Paper might create a common understanding of the human-centredness concept and give some shared vision and guidelines for how to put these basic ideas into practice. Failing that, it was thought that the paper would at least provoke engineers to react in a detailed and constructive way thereby contributing to the development of a more elaborated concept.

The shaping paper is included, in its original form, in the following section of this chapter.

2.4.2 The Social Shaping of Technology and Work
A Conceptual Framework for Research and Development Projects in the CIM area – The "Shaping Paper"

A INTRODUCTION

In this paper the concept of human-centred technology is outlined and a conceptual framework for the evaluation, design and development of human-centred CIM systems is offered. The paper is in three parts.

The first section outlines the philosophical base from which the concept of human-centredness is derived. This base, which we have termed "holism", has its historical roots in the belief that the fragmentation and simplification of industrial worklife owes much to the predominantly "mechanistic" perspective shared by technological and organizational designers throughout the industrialised world.

A critique of the mechanistic perspective is given in sections B.2 and B.3. This is followed by an outline of an alternative holistic perspective in section B.4. This section attempts to redefine the philosophical basis for understanding and changing the nature of technology and work and, as such, introduces a number of concepts with which some readers may be unfamiliar. However, concepts such as "praxis", "life-world" and "dialectic" are fundamental to a philosophy of human-centredness and have been utilized in order to avoid the use of words in common usage within the scientific and engineering disciplines which are deeply embedded in the mechanistic world view. The use of such holistic concepts, whilst running the risk of creating problems of understanding, ensures that the two mutually exclusive concepts (or paradigms) of mechanism and holism do not become confused. It would be unfortunate if "human-centredness" comes to mean all things to all men as has happened with the concept of "user friendly"!

The need for such a philosophical underpinning stems from the very nature of the human-centredness concept and allows the fundamental difference between this concept and that of "human factors", for example, to be seen. It will become clear on reading this paper that traditional human factors engineering in fact has more in common with the mechanistic world view than with human-centredness. This is because human factors engineering aims to improve the design of the human–machine interface in order to enhance performance without any consideration of the design of human work or machines in a wider context. Human-centredness, on the other hand, aims to enhance the quality of worklife and product through the fundamental redesign of work in its totality.

The second part of the paper focuses on three important dimensions of human-centred work. These are the nature of the relationship between (a) work and technology, (b) work and social communication, and (c) work and learning. These dimensions reflect the psychological importance of designing work which allows workers to use tools which extend their skills and abilities, which encourages a high degree of social interaction, and which allows the development of personality and self-determination.

The third section outlines the methods which may be used in order to apply a philosophy of human-centredness to the practice of human-centred CIM system design and development. The dimensions of work outlined in section D.3.2 and the criteria for human–machine interface design in section D.3.4 are offered as design tools to aid the shaping and evaluation of decision-making throughout the design process. They are premised on the idea that workers should be able to shape their own working practices thereby placing them in a superordinate, rather than subordinate, position vis-à-vis the technological system. The scenario in section D.2 illustrates how the system would look if these human-centred dimensions and criteria are incorporated into the design. By placing the dimensions/criteria in their overall context in this way, their usability may be considerably enhanced.

Finally, it should be stressed that this paper is not intended to be a designer's manual or guide. Rather, it offers a conceptual framework to enable social scientists and engineers to work together during the design process. Design is a complex process in which actors sharing different assumptions and interests must collaborate. When such a team is multidisciplinary, the process is even more complex and requires a large degree of mutual perspective-taking and understanding. It is hoped that this paper contributes to this endeavour.

B THE CONCEPT OF TECHNOLOGICAL SHAPING

B.1 Introduction

The concept "human-centred technology" refers to two aspects of subjectivity. First, technology shaping is a fundamental expression of human life activity. Anthropologically speaking, this ability of technology shaping has crystallized into the evolutionary process of nature that has allowed mankind to "opt out of" the process of evolution and "to enter" a social–historical process. Second, participation in the shaping of technology includes the capacity for understanding and the active participation in the shaping of one's own social conditions.

In the field of science the dimension of development and shaping is most firmly anchored in the engineering sciences and in educational theory. A related field of science includes the performance-guiding sciences, such as medicine and therapeutically oriented psychology. The purely cognitive-oriented sciences are located at the other end of the scientific spectrum. Hence, although shaping and cognition orientation are two essentially different poles of scientific study, they are, nevertheless, reciprocally constitutive for one another.

Dimensions of evaluation and analysis of work processes are also, therefore, always dimensions of the shaping of work (and technology). It must be taken into account here that designs of the future, such as those that have to be made for the planning, development and implementation of a human-centred computer-integrated manufacturing (HC-CIM) system are, in principle, full of uncertainty with respect to their effects on the experiencing process of employees, learning opportunities and stress.

Through the inclusion of qualitative methods of empirical sociological research oriented to the interpretative paradigm (see Leithäuser 1986) and to the concept of work activity research, the risk of humanely and socially "intolerable" designs of the future can be significantly reduced. In the following, a conceptual framework will be presented for the evaluation and shaping of computer-aided technology and work.

B.2 The Role of Technological Rationality

Technological rationality refers to the inherent logic contained in technical things and concepts in which the function and/or value of a technical entity or development is abstracted in terms of its usefulness. Technological functionality means that an electrical motor, for example, is constructed in such a manner that the standards of a given requirement profile (such as rotation speed, efficiency, performance, revolution–torque curve) are optimally realized.

"Optimal" means that the same technical quality can be found in the frame of a whole spectrum of possibilities, and that additional optimization criteria are required for a single solution. Technical–scientific rationality also covers the sphere of technical possibilities and the purpose-rationale of its shaping. The functioning of technology and its inner logic are, therefore, the central criteria of technological rationality.

A technological rationality assumes that a derivative connection more or less exists between technological–scientific theory, the processes based thereon, and the technical product. There are correct and incorrect steps associated with this logic. For all intents and purposes, however, there are "optimal" solutions with optimization criteria. However, concrete examples demonstrate that there is considerable shaping flexibility within the "logic". Rosenbrock points out that engineering activity has far more to do with the artistic solving of problems than generally believed. Similarly, analyses by Hellige on the significance of the technological–scientific problem-solving perspectives of technicians for the development of technology imply that the sphere of technical possibilities as the basis of technical rationality demonstrates considerable technical-cultural diversity. Thus the Scandinavian power/heat coupling approach to energy technology is fundamentally different from the condensation power plants of the USA or West Germany (Hellige 1984).

As each technology has a social context, it must lend itself more or less to social and social science scrutiny, substantiation and evaluation. Scientific theory in this respect always has a mirror function enabling a reflection of technological activity, technological process and products.

The dominant characteristic of technology is therefore the imprint made on it by conditions which, although external to it, decisively affect its concrete existence; conditions which are technically expressed, without their features being recognizable by means of an analysis of the technology. Every involvement with technology must therefore be carried out on the basis of the interest and value-oriented characteristics underlying it, even if these are difficult to discern in individual cases.

A value-free technology is a contradiction in itself, and therefore a normative understanding of technology is required. The necessity to understand the development of technology as a dimension of the shaping of social future, and to found a tradition of technology control/technology shaping, where technology is no longer regarded as an instrument for exercising dictatorial control over nature, and for the blind substitution of human abilities and activities, results from the recognition that the project does not exclude the risk of the destruction of nature and the environment.

In the dialectical relationship between nature and technology, nature is what occurs by itself. Technology has been set into motion by human beings, and refers to the world of the "artificial", that which is opposite to the "natural".

The aspect of subordination is an interchanging factor in the dialectical relationship between nature and technology. On the one hand, in the absence of mankind (which maintains and installs technology) technology withers away. Technology, on the other hand, includes the aspect of "dominance over nature", and therefore the possibility of the destruction of nature. Notwithstanding this, nature and technology are interchangeably constitutive for one another. Each technology is also based on the potentials of nature. Then again, nature is always also socially constituted nature, at least where it is in the range of human activities.

If technology in its historical and anthropological origins can still be traced to a process related to the needs and requirements from which tools and work processes developed, then it must be true also that the division of labour accompanying industrial production reflects social purposes that are abstractly represented in technology. Needs are always transmitted via interests in the process of technology shaping and the utilization of technology. However, these necessitate the political and economic structures of a society to be taken into consideration if their quality is to be discerned in the process of technology shaping and its implementation. Thus, it is only from concrete interest structures that a system of purposes develops, from which technology develops in its diverse components and interests.

The concept of technology shaping stresses the shapability of technology and stands in contrast to the concept of technological determinism, according to which technology is seen as the ultimate cause of all social movements.

Technological rationality is not identical with determinism, but the ideas of an "inner logic" and "one best way" to develop technology, which seem to be tremendously powerful, often place a strong deterministic way of reasoning by people, who are inclined to technological rationality either theoretically or in practice.*

---

* Technological determinism: . . . the first part of technological determinism is that technical change is in some sense *autonomous*, "outside" of society, literally or metaphorically. The second part is that technical change *causes* social change. Technological determinism . . . has also been employed as an historical theory, explaining why past forms of society came into being and passed away (Mackenzie and Wajcman 1985, p.4ff.).

B.3 Technological Rationality and the Mechanistic World Picture

The historical roots of technological rationality are deeply founded in the mechanistic world picture, which has been dominating Western culture so strongly in the last four hundred years. The mechanistic world picture of Cartesian–Newtonian science led to the methods of research and treatment, which are so dominating of the technological formulation as the descriptive medium, and mathematical deduction as the guiding principle, in the search for new phenomena to be verified by experimentation. It was the modes of thought inherent in this conception which penetrated into philosophical thought about the human being and his place in the universe, and into numerous special sciences which, initially, seemed to lack all contact with the study of nature.

According to the mechanistic science ideal, all subjective elements of the work process should be eliminated or objectified. Hence the declared aim of modern science is to establish a strictly detached, objective knowledge. Any theory falling short of this ideal is accepted only as a temporary imperfection, which will be eliminated later. This way of thinking is sometimes expressed in the concept of "rest-work", which means work that cannot yet be carried out by machines for either technical or economic reasons, and which therefore still requires the more or less active involvement of human beings.

The mechanistic approach perceives human beings as information processors that calculate according to rules on data, which take the form of atomic facts. Though developing still more complex and heuristic-based models of how human beings solve problems, the mechanistic oriented researchers are nonetheless convinced that by suitable questioning they can induce the expert to "recollect" the complete set of unconscious heuristics. If there is a limit to what can be understood by rules, the mechanistic approach would never recognise it. However, such an approach overlooks the fact that judgement and perception may involve global processes which cannot be understood in terms of a sequence or even a parallel set of discrete operations.

Now that the full implications of the attempt to treat man as an object or device have become more and more apparent, the mechanistic science ideal has become the target of a more serious critique from different disciplines, even from some computer scientists. The pioneers of the criticism were Heidegger and Wittgenstein. Since then, many others – notably Maurice Merleau-Ponty, Michael Polanyi and Hubert L. Dreyfus – have, each in their own way, applied and refined similar insights.

The mechanistic science ideal seems to have striking parallels in the more practically oriented management circles and system development in relation to information technology. Weltz and Lullies (1983) have summarized some characteristics of the mechanistic image of the human being, based on numerous interviews with managers and system experts. They distinguish between two different images, both related to the mechanistic approach. The most rigid of these considers the human being as a disturbing factor to be reduced through extension and perfection of the technical system (the vision of the fully automated factory). The more flexible but still mechanistic image considers human work as a mountain river, whose energy is to be channelled through organizational rules and opportunities of control. These two views of the human being imply strategies of rationalization, which: exclude all "negative" possibilities; prescribe the "correct" behaviour as precisely as possible; restrict the human initiative as much as possible; and develop still more methods of possible control. Weltz and Lullies further argue that this process of rationalization, which has dominated the historical development of work in industrial society, has produced a tendency to passify and alienate at least some employees.

Historically, this alienation relates to the threefold process of subdivision of work, mechanization and automation, each of which fed upon and reinforced the others. Before jobs can be taken over by a machine, they must first have been fragmented. This is almost always achieved in developmental stages; at each stage the jobs given to men and women are whatever remains once the jobs of the machine have been determined. Therefore, subdivison of work was a precondition for mechanization and automation, which incorporated still more human movements and judgements. Part of the result of this development has been the shaping of a large number of trivial repetitive tasks, and the creation of a workforce who are rendered more passive as technology becomes more active. In the mechanistic view, this passivity of employees reinforces the view of human beings as solely motivated by material rewards (e.g. wages). Such an interpretation may stem more from a psychological reaction to trivial work than from any innate trait.

The emergence of a counter-tendency to this rigid perception is becoming more and more clear, for both functional and human reasons. The functional critique, which is observable in some management circles, focuses mainly on the lack of flexibility and responsibility with respect to sudden changes or "unforeseen" situations. Thus the rigid perception of human beings may result in counter-productive effects, which force management either to increase the degree of external control or to change the mechanistic perception of the employees. The human critique is based on a perception of the human–work relationship quite different from the pessimistic, mechanistic image mentioned above; this different vision is presented in the next section.

B.4 Work and Personality from a Holistic Perspective

B.4.1 The Search for a New Paradigm

The critique presented in the previous section of the mechanistic approach is rather fundamental. It is not just a specific explanation, but the whole conceptual framework that has failed.

Ever more scientists from different disciplines – including physics, biology, psychology, sociology – seem to be searching for a holistic paradigm which transcends current disciplines and conceptional boundaries. As Fritjof Capra points out: " . . . We live today in a globally interconnected world in which biological, psychological, social and environmental phenomena are all interdependent. To describe this world appropriately we need an ecological perspective which the Cartesian world picture does not offer" (Capra 1982, p.XVIII).

So far most of the scientists from the above-mentioned disciplines still operate independently and have not yet recognized how their intentions interrelate. Nevertheless, an important shift of paradigm seems to have developed in many scientific disciplines for both materialistic and epistemological reasons.

B.4.2 The Concept of "Life-World" and "System"

The concept of life-world (Lebenswelt) is defined by Habermas as the "praxis relation" of the human community. It is the horizon of meaning with the content of more or less tacit knowledge, which helps social actors interpret the different situations of action. The life-world functions both as an implicit horizon or context (and thereby partly at the unconscious level) and as resource of convictions, etc., which the human actors use to establish consensual or conflicting interpretations of specific situations. The communicating actors are placed in a context of specific situation (e.g. work situation) and in a context of life-world. The situation determines some flexible restrictions and the life-world functions more indirectly as both a horizon and resource for the interpretation of actions.

Following Habermas, the structural components of the life-world may be analysed from three partly interdependent aspects (Habermas, 1970):

1. Culture (mutual understanding, traditional and new knowledge, meaning).
2. Society (coordination of actions, norms of legitimacy, social solidarity and integration).
3. Personality (socialization of the individual person with different sorts of competences (cognitive, moral, expressive, etc.)).

The reproduction of the life-world has both a material and symbolic dimension. The relation between the life-world and situation-based communication may be characterized as reciprocal action and thereby contains the possibility to change part of the life-world. The system is a formalized or institutionalized part of the life-world based on money and power.

The two essential characteristics of the evolution process of society are: (a) the increasing rationalization of the life-world and (b) the increasing complexity of the system differentiation. Rationalization of the life-world results in an increasing differentiation between culture, society and personality. In the modern society this tendency has produced an institutionalization of expert cultures, which develops the three structural components of the life-world from each specific demand of validity.

The increasing system complexity is a historical, developmental process taking place at many different levels of society (including the internal structure of the firm). It is in some way reduced by the establishment of new subsystems, which then force the problem of integration. But integration can be either functionally or socially oriented. Social integration can be seen as part of the symbolic reproduction of the life-world with the purpose of maintaining or increasing the system perspective.

According to Habermas, the system colonizes additional areas of the life-world. In a capitalist society the consequence is often disintegration of the traditional modes of life without establishing new meaningful ones. However, the development is not exclusively negative; it may provoke a tendency to search for new, alternative ways of living.

B.4.3 The Concept of "Praxis"

As mentioned, the concept of praxis is closely interrelated with the concept of life-world. It is going back to the ancient Greek philosophers thinking on the difference of theory and praxis. Praxis in this sense means something different from practical or technical work. On the contrary, one has to look at human activities from the point of viewing them as being imbedded in situations, where the moral question of "what is to be done" is at stake. That is, the question of "how it is to be done", should be answered within the frame of the complexity of human life as a whole. There exist very different fields of praxis, such as policy, science, industrial production and individual life. From a holistic point of view, the purpose cannot be a simple reintegration of these different fields of praxis. But inside the different fields of praxis it is necessary to reflect the conditions, validity and consequences of the special praxis. When doing this, the special praxis is always confronted with the overall interrelatedness of culture, society and personality. In science, as well as in other fields of praxis, it means that without such reflections it cannot be a humane praxis.

Both individuals and social organizations contain the potential for self-renewal and self-transcendence. Self-renewal refers to the ability of organizations and organisms continuously to renew and recycle their components, while maintaining the integrity of their overall structure. Self-transcendence refers to the ability to reach out creatively beyond physical and mental boundaries in the processes of learning and development.

If the potential for self-renewal and self-transcendence are prioritized in the shaping of the prospective work organization, the mechanistic picture of workers as mechanical performers of orders must be abandoned. Instead, workers are to be considered as purposeful actors, capable, when possible, of performing meaningful acts based on their own subjective knowledge and judgement.

Moreover, the objectivist paradigm, which considers data as a mapping of reality in a context-free manner, has to be abandoned. Communication through language or non-verbal gesticulation is an act of commitment and interpretation based on a more or less common social and historical background.

Hence, the concept of praxis developed above relies on two sorts of interdependences:

1. The mutal interdependence and connection between body and mind, including the conscious and unconscious levels.
2. The dynamic relationship between work conditions and the psychosomatic processes of the workers involved.

B.4.4 The Relation Between Mind and Body

The mechanistic way of thinking – following the tradition of Plato and Descartes – has viewed the body "as getting in the way of intelligence and reason, rather than being in any way indispensable for it" (Dreyfus 1979; p.235).

In the holistic approach the body–mind relationship is considered crucial to an understanding of how skilled workers and designers develop and use their tradition-based abilities in practice. One essential point is that an embodied skill, unlike a fixed response or set of responses, can be brought to bear in an indefinite number of ways. Focusing, getting the right perspective, and picking out certain details, all involve coordinated relationships between mind and body.

Dreyfus concludes that the body contributes three functions not present, and not yet conceived, in digital computer programs: (a) the inner horizon, i.e. the partially indeterminate, predelineated anticipation of partially indeterminate data; (b) the global character of this anticipation which determines the meaning of the details it assimilates and is determined by them; (c) the transferability of this anticipation from one sense modality and one organ of action to another.

All these are included in the general human ability to acquire bodily skills. Thanks to this fundamental ability, an embodied agent can dwell in the world in such a way as to avoid the infinite task of formalizing everything (Dreyfus, 1979, p.255).

It seems essential to find ways of restoring or improving the body–mind relationship, even when a CIM system – or part of it – is developed and/or implemented. From a holistic perspective it may be fruitful to consider the mind as a complementary dynamic process between the conscious and unconscious levels, viewed as a continuous mutual stream of energy. The unconscious level is not only a dump for repressed instincts, but may also be seen as a creative and intelligent principle relating the person to the natural, social and biographical background.

Polanyi distinguishes between two kinds of mutually exclusive awareness (Polanyi 1967): focal awareness, and form awareness. This means that a person may be aware of certain things, either by focusing on them directly, or by looking at them as functional parts of a superior target. Focal awareness always is fully conscious, but form awareness may exist at any of the conscious and unconscious levels. Polanyi's ultimate focal aim is that all knowledge is either tacit or rooted in tacit knowledge. Besides, he argues, every concept's relation to reality is grounded in personal commitments, which we are unable to specify completely. The reason is that we are dwelling in these basic commitments. Focusing our attention upon them means destroying their subsidiary function. Therefore, we cannot spell out this process of tacit integration in explicit steps.

Winograd and Flores (1986) also stress the importance of participation and involvement rather than formal operations: "Knowledge and understanding (in both the cognitive and linguistic senses) do not result from formal operations on mental representations of an objectively existing world. Rather they arise from the individual's committed participation in mutually oriented patterns of behaviour that are embedded in a socially shared background of concerns, actions and beliefs."

In their book *Mind over Machine* Hubert Dreyfus and his brother Stuart Dreyfus try to describe how analysis and intuition work together in the human mind at different stages of the skill-acquisition process (Dreyfus and Dreyfus 1986). Their essential point is to show that analytic thinking and intuition are not two mentally exclusive ways of understanding or making judgements. Rather, they seem to be complementary factors working together, though intuition is of increasing importance when the skill performer becomes more and more experienced. With enough experience, the human being seems to be able to recognize whole scenes without decomposing them into specific features.

On the other hand, one might have to take into account that, nevertheless, there are important distinctions between analytical and intuitive ways of understanding and reasoning. When it comes to the point of integrating life-world into every-day working inside the factory, especially, it should be necessary to realize that people are interested not only in their present work, but in other fields of life-world as well, and at the same time. In a "human" concept of work, there should be opportunities of leaving space opened for both ways of approaching the "world", fulfilling actual tasks as well as having a chat with one's friend about something completely different.

Both the design function, and skilled machining, are examples of occupations characterized by tacit skills. However, the choice is not between tacit or calculative skills, but rather how to find the balance between analytic thinking and intuition. Both are necessary from the holistic point of view of the human–work relationship.

B.4.5 The Relationship Between Working Conditions and the Personality

In the holistic approach, personality and work are dialectically related to one another. This relationship has two aspects:

1. Personality-forming work aspects
2. Work-related personality aspects

The basic model of the relationship between work and personality, insofar as it concerns the work sphere, is structured according to degrees of restriction, and constellations, whilst the personality aspect is structured according to its flexibility or rigidity. Here, flexibility signifies conscious activity, whereas rigidity means conduct.

The reorientation towards a positive theory of the shaping of technology and work, as it has been attempted by various social science studies over the last few years, is as difficult a process as the increasing attempts within engineering and natural sciences to come to terms with the social implications of technology in the technology-shaping process.

The reorientation towards a positive theory of technology (and work shaping is not only a matter of crossing the borders between traditional separated scientific disciplines) also necessitates the integration of the theoretical and practical knowledge and experiences too. The possible users of the prospective technology are to be involved in the shaping process, not only at the stage of implementation and adaptation, but also in the initial development process. This reorientation includes a number of different fields in society, in which technology shaping is influenced. In the following section the scope of these fields is tentatively presented.

B.5 Fields of Technology Shaping

Concepts of technology shaping are usually limited to specific social sites of technology shaping and thereby neglect the interrelationships between the social sites. Figure 2.7 represents an attempt to view technology shaping in a more holistic way. On the vertical axis of the figure, the social sites of technology shaping are disaggregated into different categories (ranging from the individual through to international socio-economic and political structures). The different spheres or sectors of technology (e.g. electronic, mechanical, chemical process and nuclear technologies) and their application are represented on the horizontal axis. The figure therefore allows an assessment of the importance of individual approaches to the shaping of technology (relative to other specific social sites) to be made across a range of technological applications.

This contextual approach to technology shaping raises the question of what significance the ability of shaping activities located within industrial vocational education and training has for the more macro sites of technology shaping (Braten 1984).

Individuals are differentially affected at the different sites of technology shaping; for instance, at the State level as a voter, at the university social site, as an expert, as a specialist, student, etc. Each concrete individual is alwaysboth actively and passively involved and affected, as, for example, with wage dependence, as a citizen and as a private person and consumer, and he or she is frequently faced with conflicting interests within his or her own self. The total personality must always be considered.

The evaluation of a technology can lead to vastly different interpretations according to the perspective chosen. Thus, from the perspective of a member of a factory or enterprise, the evaluation of a technology can be a very different interpretation than from the viewpoint

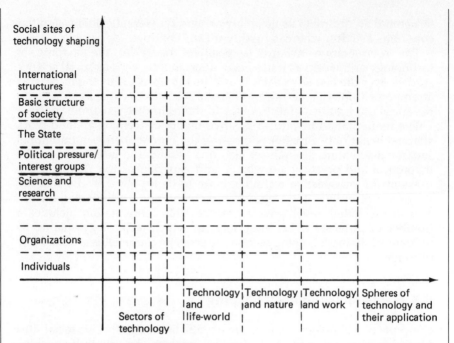

Fig. 2.7. Fields of technology shaping.

of a member of a social movement. For example, the education/training concept "technology shaping" requires the integration of the perspective of the affected, as well as a differentiation according to sectors of technology. Every educational/training concept, but in particular all occupational education/training, would be inadequate if it were to reduce the personality to the "dimension" of skilled worker. This differentiation is necessary from the perspective of shapability and shaping of technology, as the weighting of the shaping fields and their relationship to each other is very different according to the different areas of technology.

The development of the concept of technology shaping within the "sphere of technology and its application" necessitates the arrangement of the areas of technology at one end of a horizontal scale, emphasizing the basic innovation, the fundamental technological–scientific core of this technology. With regard to micro-electronics, the basic innovation is the integrated programmable electronics system, as represented inter alia by microprocessors. Those micro-electronic- or microcomputer-centred products and processes directly applicable/implementable can be classified at the other end of the scale.

Thus a scale develops according to which every technology/technique which can be categorized in the sphere of micro-electronics can be classified between the poles of "fundamental basic innovation" and "concrete application of the technology".

This differentiation is meaningful for the shaping question, as the purpose–means relationship continually moves along this scale. Should the indirectly employable technology, such as very specific user software, still be able to be easily extrapolated from the previously defined purpose, which is quite unequivocally defined in the form of an operationalized manual of duties concerning the product to be developed, then the purpose–means relationship at the other end of the scale – "micro-electronics as basis innovation" – can be reversed.

Micro-electronics seeks its own purposes. Only in this sense can we speak of an open-purpose structure in relation to communication/data technology. The opinion that the crystallization of open-purpose structures to the development of technology is a phenomenon of the most current technological developments cannot be substantiated. Every sector of technology can in principle always be ordered along a scale that allows differentiation on the basis of that technology and its concrete use.

Then, however, an implicit displacement of the purpose–means relationship along the continuum between technical basis innovation and its multifarious applications also holds true. In this sense, the building blocks of micro-electronics are nothing more than the new (and old) basic building materials, as is the case with plywood in the woodworking/timber industry, or bricks in the building industry. The difference between the one and the other area of technology does not lie in this fundamental phenomenon, but rather in the shaping potential inherent in a technology, and the opportunities for direct and indirect involvement in the technology shaping process.

The significance of the reversal of the purpose–means relationship to a means–purpose relationship is exemplified by transistor technology. Transistor technology was developed by Bell Laboratories in the frame of the management goal of creating a solid-matter triode to replace the vacuum valve in order to be more competitive in the amplifier technology market. Transistor technology was to become the basis for innovations in microcomputer technology. Yet this was neither formulated as a goal, nor was the imagination of those involved at the time sufficiently flexible to anticipate even the slightest hint of such future development possibilities. Hence, this problem of the open-purpose structure of basic technical innovation must be considered in the debates about humane and socially compatible shaping of technology.

In fundamental projects for the further development of engineering and natural science research, the capacity for technological and socially creative imagination on the part of the researchers must be taken into consideration in any reappraisal of possible historical ramifications and for responsible participation in technology shaping for the future. This creativity is essential in order to anticipate, and reflect upon, future development phases. Added scope can be provided by the formulation and design of well-founded scenarios which articulate possible future developments and uses of research knowledge, products and their possible effects and consequences.

Yet, whilst the general recognition that technology is an expression of social purposes (as well as the interests underlying such purposes) is necessary if we are to develop a concept of technology shaping based on a specific technology, this is not sufficient. Even if we were to succeed in realizing a use-value oriented concept of technology development for the purpose of creative and gainful employment, one key problem remains. History repeatedly shows us that the design and development of technology can have social and ecological effects and consequences which are both unintended and unforeseen by its designers, developers and users. A solution to this problem in the sense of technology shaping, which applies especially to the technology developed according to the analysis–synthesis concept (non-holistic, non-ecological), includes the abandoning of socially non-controllable technology.

The extent of the side-effects of technical products and processes as a rule exceeds the dimensions of the intended consequences of purpose-rational technological developments. In this respect, technology is in principle always also incalculable. It bears surprises and unforeseeable potential and gives rise to previously incalculable consequences.

A concept of technology shaping, and an education/training goal allied to the shaping of technology, must therefore take account of both aspects of technology shaping:

1. A use-value-orientated shaping of technology in unison with qualified economics.
2. A shaping of technology that is aware of the limits of purpose-rational technology development and the incalculability of the consequences of technology and/or its innovative potential.

C DIMENSIONS OF HUMAN-CENTRED WORK

C.1 Introduction

The development of a human-centred CIM system presupposes that this CIM system is necessarily regarded as a technology facilitating the establishment of preconditions and (work) conditions, whose "humane character" must first be proven in the concrete work process. Hence the HC-CIM technology that is to be developed has a hypothetical character.

The justification, substantiation and evolution of a concept of "humane work" should differentiate according to three work-oriented relationships: (a) work and technology; (b) work and communication; (c) work and learning.

C.2 The Relationship of Work and Technology

In Western culture there exists an analytically explicable connection between work and technology, inasmuch as the knowledge implemented in the forces of production, embodied in machines, organizations and competencies, is meant to improve the productivity of labour by improving the control over nature and over cooperating human beings.

As mentioned earlier, a dramatic transformation in the relationship of work and technology has occurred during the course of the Industrial Revolution. This is characterized by the relationship of work to technology (as a measure in the work process) which was/is increasingly organized from the perspective of machines and "rest-work". This ongoing process of substitution and objectification of work presupposes a process of increasing regimentation and division of more complex work procedures that need to be carried out.

The work organization serves as a mediation instrument for the acceleration of the segmentation and objectification process. This development accompanied the mounting process of subordination of the human being to the machine. Thinking in terms of machines became the norm for the shaping of the work–technology relationship. In the resultant competitive situation of man versus machine, the machine usually proved to be superior in the work process. If one defines and organizes human abilities as mechanical, insofar as humans are required to carry out robot-like tasks, then the "proof" that a mechanical robot is superior to the "human robot" is usually close at hand. Such a logic is now increasingly being applied to office automation (Bailey 1982).

We can formulate this in general terms. If one always defines and organizes work in the work–technology relationship from a machine perspective, then one makes the machine and its "abilities" (functions and possibilities) the norm for human abilities. The argument that industrial technological development has increasingly freed us from routine work and stress burdens is only partially true, as:

1. It is often overlooked that burdensome work frequently derives from the same logic as the technology that was and is implemented for the segmentation and objectification of that work.

2. A very restricted mechanistic comprehension of stress burdens underlies this form of thinking.

Where work is experienced as hardship, and generally regarded as burdensome, all forms of objectification of work appear as a reduction of stress, and a "freeing" from work would ultimately correspond to a burden-free situation. "Social work" is, however, not only objectively necessary for survival, but in the shape of human work is subjectively an intrinsic prerequisite for human existence. If one does not regard work, from the outset, as wage labour, then it becomes apparent that cause enough exists for understanding work as a primary form of human life expression. Moreover, this awareness ought to lead to the shaping of work accordingly, and not allow it to atrophy to a mere reflection of technological development that excludes the human being from (collective) productivity.

The human bioogical development potential, as it is biologically and historically given, must become a point of departure for the shaping of work and technology. The potentialities of self-renewal and self-transcendence – both at the individual and organizational level – are given high priority from the human-centred perspective. This includes the capacity for understanding and the instance of active participation in the shaping of one's own social conditions.

Thus, the development of "human-centred" computer-assisted work and technology, particularly in the frame of research and development projects oriented towards an integrative technology-shaping concept, should include the following:

1. Analysis of the formative conditions and conditions of change for computer-assisted and computer-integrated production technology, including possible ramification opportunities.
2. The historical reconstruction (re-evaluation) of alternative development directions in the domain of computer-assisted work and technology.
3. The analysis and assessment (evaluation of the relationship of computer-assisted technology) and the social interests of those involved and concerned (internal) groups and individuals, as well as the subjective experience of the employed.
4. An analysis and evaluation of CIM under special consideration of the diverging interests, as they are given via the diverse social sites of technology shaping and the individuals acting at those sites.
5. The development of tables of duties for the shaping of HC-CIM components, in which the social implications and intentions are apparent, and the shaping-scientific possibilities in which shaping criteria for industrial work and life-world are formulated.
6. A model realization and testing of the HC-CIM system, since only by means of the real work process can its quality more or less be proven to be human-centred or not. In other words, only via a longer-term development process of HC-CIM work can the intended development possibilities be brought to fruition.

C.3 Work and Communication

Communication is a fundamental human relation (Holzkamp 1973). In the speech act theory developed by Habermas the act of language and communication is emphasized rather than its descriptive or representational role. Communication through language or non-verbal gesticulation is an act of commitment and interpretation based on a more or less common social and historical background. The act of communication is directed towards the creation of mutual orientation (Maturana). As pointed out by Winograd and Flores, this orientation is not grounded in a correspondence between language and the world, but exists as a consensual domain – as interlinked patterns of activity (Winograd and Flores 1986).

Communication, as defined here, influences the work processes more deeply than is reflected in the perception of the mechanistic approach, which emphasizes the role of transmitting information and symbols. Through the communicative act the actors not only give or obey orders in the work process, they also confront or share interpretations of experiences regarding the social and functional aspects of the common work-life.

As pointed out by Leithäuser, there exists a "hidden situation" in every firm under and beside the formal organization of the firm, which is largely unrecognised by the management or the external system experts. The centre of the "hidden situation" may shift or take different shapes dependent on how extensive and advanced the technology, but we may anticipate that even the most integrated and complex system is still dependent on the interactions taking place in hidden situations.

The communicative acts in the hidden situation often supply or rectify the direction of the work process, when the officially prescribed task or methods are inadequate. Moreover, it is important to restore the hidden situation regarding subjective self-respect and human dignity. Therefore it is not surprising that even workers at terminals with restricted social contacts have developed an informal culture through the communication network.

Instead of trying to eliminate the possibilities of informal interaction – which seems to be the dominating tendency in conventional technology shaping – a holistic technology shaping may deliberate how to widen possibilities of informal interaction, both in the single work-group and between different departments.

As pointed out by Winograd and Flores, the importance of face-to-face communication has traditionally been preferred by successful managers (Winograd and Flores 1986, p.151):

> . . . their activities are not well represented by the stereotype of a reflecting solitary mind studying complex alternatives. Instead, managers appear to be absorbed in many short interactions, most of them lasting between two and twenty minutes. They manifest a great preference for oral communication – by telephone or face to face.

But this may be the preference of other parts in the firm too. Designers and skilled workers may prefer just as much oral communication as managers.

In the context of computer systems, the predominant emphasis is on formalized information and preplanned communication. The significance of non-formalized information and communication is often neglected, perhaps due to the difficulties of making it visible. Many systems designed by computer professionals are intended to facilitate the activity of an individual working alone, and thereby leave out the essential dimension of collective work. The introduction of integrated systems with time-sharing terminals has often radically cut off natural face-to-face communication channels. From both a functional and human perspective such tendencies could be counteracted. A holistic technology shaping may deliberate how to widen informal and face-to-face communication as possibilities both in the single work-group and between different departments.

Communication mediated by computer applications may constitute a supplement to face-to-face communication, but not a substitute for it. Thus, formalized and preplanned communication may readily be interpreted as a part of overall communication. If there are any problems in technical communication, it may be shifted to personal face-to-face communication. Furthermore, if the objectives of coordination between two or more collaborating persons are known, the choice of communication medium may be based on their own decision.

As pointed out by the Norwegian sociologist Braten (1984), decentralized computer networks may increase symbolic centralization. If a "model-strong" actor and a "model-weak" actor are coupled in an open information exchange system, the former may be expected to gradually increase his control of the other actor:

> Thus a successful transition in the name of democratization to a more open communication structure may freeze – or even increase – the influence gap . . . In order to restore balance, and thereby dialogue in such systems of interaction, it is required that the participants are able to step outside the system's boundary, as it were, and assume an observer's position. (Braten 1984, p.190)

Communicative action should be rationalized neither under the technical aspects of the means selected nor under the strategic aspect of the selection of means. The principle in the holistic approach should be the moral–practical aspect of the responsibility of the acting subject and the justifiability of the action norm (Habermas 1970).

To do this is not only organizational but also an educational matter:

> There exists a domain for education in communicative competence . . . that is the capacity to express one's intentions and take responsibilities in the network of commitments that utterances and their interpretations bring to the world . . . People's conscious knowledge of their participation in the network of commitment can be reinforced and developed, improving their capacity to act in the domain of language. (Winograd and Flores 1986, p. 162)

Thus we touch on the third dimension of human-centred work: the learning aspect, which is discussed in the following section.

C.4 Work and Learning

Work, from an anthropological perspective, is always an acquisition or learning process. During the work process, the expenditure and acquisition of abilities always takes place as two dimensions of the same process. Hence the volume of learning opportunities within the work process is a measure for humane work and technology. In the frequently horizontally and vertically divided work process, the chances for learning (particularly in the range of simple performing activities) are restricted to a large degree.

In the search for new production concepts in the area of computer-assisted work and technology, this insight, as well as the recognition that a phasing out of the acquisition aspect of work is counter-productive, is beginning to find acceptance. The reintegration of previously fragmented work and the extension of performance flexibility as instances of rehumanization of work (combined with a strengthening of the quality of acquisition (learning opportunities), motivation, and identification with the work) are becoming sought-after productive forces in computer-integrated production.

The production island is thus increasingly gaining in significance as an efficient production concept. "Production islands", with their homogeneous qualification structure, can, from a manufacturing as well as a time perspective, react better to the varying requirement of the market than centralized and hierarchical forms of organization. Following extensive investigations of small and medium-sized enterprises, Moll (1983) comes to the conclusion that "small craftsman workshops" (Kleinmeistereien), with "flat" hierarchies and elastic cooperative relations, should be essential elements of future production. What is needed are highly and broadly qualified skilled workers willing and able to learn, with market company loyalty and professional qualifications. An undifferentiated extrapolation of the uninterrupted Taylorization of work until the late 1970s is unacceptable.

C.4.1 The Relationship Between Cultivation (Bildung) and Qualification

Qualification refers to a spectrum of requirements of skills and abilities demanded of an employee or group of employees, comprised more or less of the specific content of work and work forms. Qualifications are therefore also always profiles of activity requirements relatively external to the personality of the working individual. The attempt to shape the activity requirements, as they crystallize according to the concept of "rest-work", into a personality theory in which categories of activities are placed in antithesis to a typology of psychological dispositions (from which a hierarchically organized automatic control system model is then postulated as a personality model) is therefore unsuited to explain the relationship between work and personality.

Seve referred to the misplaced attempts at "importing" models from mathematical logic, cybernetics and linguistics into the psychology of personality, as "linguistomania" and "cybernetomania". An example of a psychological theory that can be classified in the category is the "performance-regulation theory" as formulated by Hacker (see Simon 1978).

The concept of qualification refers to the qualification holder. This holder does not become subject to the cognition and shaping processes. Qualification bearers appear as a residual risk from the perspective of the primary qualification requirements, as interference potential and a source of unpredictability. Therefore they are variables that could be disposed of. The substitution of human representatives by mechanical representatives is a distinguishing mark of mechanization and automation. A work process can be described by qualifications. Subjective elements of personality, however, cannot. Qualification is an economic category. Education, on the other hand, refers to personality in the relationship of work and personality. From the cultivation (Bildung) perspective, the question is raised about the cultivation value of work content. The unfolding of the personality forces on the basis of an intellect capable of self-determination is the starting point of the education concept as well as of practical cultivation efforts.

A greater contrast of cultivation and qualification leads to unacceptable reductions in the description of the fluctuating relationship between work and personality.We therefore base our approach on a dialectical relationship between cultivation and qualification. Cultivation and qualification are, as depicted above, substantially different from one another, and in this sense on opposite sides of the fence, i.e. they represent the poles of a dialectical relationship.

Cultivation is therefore a constituent aspect of qualification since it always takes place in an insoluble alternating relationship between society and personality. At the same time, qualification always contains cultivation aspects; qualification is implicit cultivation. The suppression of this alternating relationship in the practice of cultivation and work planning leads to violation of the personality on the one hand, as well as (where applicable) to inhuman shaping of work, and production processes on the other hand. Qualification conditions and opportunities should thus be seen as prerequisites for educational opportunities, as well as vice versa. In a project with the claim to being human-centred, the primary perspective in regard to the work process dimension is the cultivation perspective. Qualification requirements need to be developed, from the perspective of cultivation interests and opportunities. For the alternating relationship between working/learning in a computer-assisted production process, this means that besides the procuring of instrumental qualifications (trade, craft, intellectual, emotional) the opportunity to comprehend the work process must be available, both in this technical–economic connection (shop-floor and external), and from an ecological and humane, as well as social–historical perspective, and to reflect upon these.

As work processes cannot and should not be counteracted or neutralized in cultivation processes – they are rather different processes – it is necessary to develop models according to which the sketched cultivation values can be realized. The place of work, as an implicit centre of cultivation, is of great significance in this respect. Cultivation and qualification, however, cannot be limited to the work process. The cultivation goal of participation in the shaping of work and technology, as a central idea for the project human-centred CIM, presupposes a professional education concept far broader than the perspective of the work process. Yet such breadth is essential if we are adequately to unfold and shape the interrelationship between humane computer-assisted skilled work and qualification.

More humane computer-assisted manufacturing which enables greater performance flexibility (autonomy), more opportunities for learning (qualification), and the realization of less stress, can, when coupled with the dismantling of "indirect" skills in separate planning and steering spheres, lead to an increase in productivity (Moll 1979).

The dynamic productivity model, proposed by Rosenbrock and developed further by Seliger, demonstrates that productivity gains, if they are directly transformed into short-term rationalization gains, can lead to unfavourable economic consequences in the long term. Given this, and considering the discrepancy between education/training and qualification, this model may be useful for the implementation of an HC-CIM system (Rosenbrock 1981; Seliger 1983).

Besides the utilization of opportunities for learning in the work process, the anticipated growth in productivity is translated into further industrial qualification measures. This further education/training is, above all, directed towards the transmission of creative shaping, as well as social and reflective abilities.* As the discrepancy between education/ training and qualification also includes the discrepant relationship between individual and collective interest in education/training of the employees on the one hand, and the interest of the management in the utilization of qualifications on the other, it would be of fundamental value to shape the concept of further education/training in such a way that the right of education and interest in education of the employees, and the qualification interest of the firm, are sufficiently taken into consideration.

D METHODS FOR EVALUATING AND SHAPING COMPUTER-AIDED TECHNOLOGY AND WORK

D.1 Introduction

The mechanistic approach to the design and development of computer-aided technology and work typically follows a sequential process in which the technical aspects of a work system are designed first, and psychological and organizational aspects are not considered until the system is implemented. The technical system is thereby shaped by the mechanistic science ideal, and produces demotivated and passive behaviour amongst the users of the system (as discussed in section B.2).

A holistic approach to the design of computer-aided technology and work must involve the consideration of human-centred technical and social criteria from the beginning of the design process. Amongst engineering designers, the design of technology and work is viewed almost solely as a technical concern and it is therefore important that some method whereby human-centred considerations can reshape this process is made available to designers, if this trend is to be redirected.

It is possible to distinguish five distinct, although essentially complementary, methods of parallel design in the social science literature. These methods are briefly discussed below and the advantages and disadvantages of each are summarized in Fig. 2.8. A more detailed discussion is included in Ravden et al. (1987).

Sociotechnical systems design developed out of the work of Fred Emery and Eric Trist at the Tavistock Institute in Great Britain. The method associated with the sociotechnical approach to work design is the nine-step variance analysis technique. This method is extremely complex (and time-consuming to undertake) as it includes the analysis of user and supply systems and the analysis of market environment and

---

*Brater in his differentiations concerning the limits of a vocational education/training concept oriented towards a purpose-rational approach, comes to the conclusion "that the particular requirements and expectations demanded of skilled workers thus exist above all especially therein, of being able to *master complex, not standardised and relatively diverse tasks flexibly*, as in small series production or tool-making" (p.65).

organizational development plans. This macro-level of analysis is reflected in the almost exclusive application of sociotechnical design methods in the design or redesign of complete departments or factory-wide production systems.

With a few notable exceptions (e.g. Volvo Kalmar), most of the work in this field has concentrated primarily on the redesign of existing social systems, with little or no redesign of the work technology itself. Indeed, it has been argued that the methods of sociotechnical design are largely mechanistic in practice since the design of operator tasks are based on the control of instability arising from the technical system. Hence, sociotechnical design tends to be technically led.

The parallel design method of user participation has gained increasing prominence in recent years. The rationale behind this method is that if system users are involved in design, then the resultant system will be better, either because it will suit the particular needs and skills of those

Design method	Advantages	Disadvantages
SOCIOTECHNICAL SYSTEMS DESIGN	Comprehensive. Can be used to design complete factory	Very complex – requires sociotechnical systems specialists. Tends to treat technology as a "given" in design. Aids design choices only at a general level
USER PARTICIPATION	Caters for specific user needs. Uses a full range of personnel expertise	Unfamiliarity of technical details and design options to users may make participation difficult or ineffective. User involvement tends to lag behind technical design. User group may not be identifiable or available. Tends to produce "conservative" systems
DESIGN CRITERIA AND GUIDELINES	Can be very specific. Allows design choices to be evaluated at an early stage	Less suitable at general level. Limited knowledge base. May be difficult to use
SCENARIOS	Relatively easy to construct. Inexpensive	Unsuitable as aid to specific design decisions. Leaves design process technically driven
DESIGN BY DOING	Exposes design options to users, allowing more informed user contribution. Unconstrained by conventional engineering design custom and practice.	Very time consuming. Essentially experimental and therefore difficult to manage and may produce unrealizable system specification

Fig. 2.8. Comparative advantages and disadvantages of parallel design methods.

people working with it, or because these people have the job-specific knowledge about the production process which should be included in the technical design itself.

The actual form which user participation takes in a particular design project may vary, although the most successful projects have involved users during the entire design and development process. There are a number of problems associated with this method and these are summarized in Fig. 2.8.

A third method of parallel design involves the representation of social science expertise in the form of design guidelines or criteria for use by design engineers. The aim of this approach is the inclusion of these "social" criteria in the system, alongside engineering and economic criteria, in order to optimize the human and social aspects of the system. The major problem associated with this method is that of communicating the knowledge base behind the criteria in a form that is usable by designers (Corbett 1987).

The use of scenarios – a fourth parallel design method – is a recent development and it is difficult to evaluate its effectiveness in shaping design. A scenario is basically a picture or vision to illustrate what a system would look like if certain social considerations were included in the design. By itself, a scenario is only a starting point for parallel design, and may leave the design process largely unaltered, and therefore technically driven.

The most radical method of parallel design is "designing by doing". This method derives from the work, in Scandinavia, within the UTOPIA project. It is a participative methodology involving users from the very onset of design. Users are able to articulate their needs in a concrete way, by using prototypes and mock-up simulations, and then draw up a specification for use by design engineers. This experimental approach to design is still under development and, although a number of problems associated with user participation may be overcome by this method, communication between users and designers remains problematic. The technically unconstrained nature of the user involvement is difficult to manage and may lead to unrealizable system specifications unless steps are taken to involve design engineers more closely in the early stages.

In this final section, a strategy of social shaping is outlined which attempts to combine three of these parallel design methods in order to minimize the disadvantages associated with the individual methods. These methods are: the use of scenarios, design guidelines or criteria, and user participation.

The focus of this strategy is the use of the six work dimensions (criteria) discussed in section D.3.2. These are:

1. Time structure
2. Space for movement
3. Social relations
4. Control flexibility
5. Qualification
6. Stress control

One important way by which these dimensions may be used during the design process is through the drawing up of a scenario, or picture, of the ideal typical work environment. Such a scenario enables designers to see the kind of work environment into which they must project their end product, and allows design choices to be evaluated according to how well they contribute to such a factory work-life.

D.2 A Scenario

D.2.1 Range of Products

The factory of the future will focus on small to medium batch production and contain flexible production islands incorporating both highly automated and manual-intensive machines. Flexibility here is taken to mean the ability to adapt to changes in product design, batch size, and machining processes in order to meet changing market demands and customer requirements. To meet these demands, the human-centred technology must be integrated with, and complemented by, an equally flexible organization of work. We are therefore talking about human-centred sociotechnical CIM systems.

The factory of the future produces a wide range of products with a certain common core. These products will include "fixed" products presented to customers in catalogues, customer-specified products and products adapted to customer requirements. In all cases the factory will design and manufacture the complete product without the use of a subcontractor. Relationships with customers are direct to allow the evaluation and improvement of the use-value of the products. In this way the factory can develop its product range in new directions as expertise in design and manufacture (and customer requirements) develop.

D.2.2 The Factory

The factory is made of three sectors: coordination, design and manufacture, and these are organized by product and not by function to maximize flexibility. In other words, instead of dividing labour, work is organized by dividing products and product order. This is achieved by extending the concept of group technology into a general organization concept rather than a single technique.

The factory's product range is divided into part families, i.e. groups of products with similar characteristics, and the production islands are responsible for a particular subgroup of these part families. The production islands have the task of producing components and complete products as far as possible from raw materials. All necessary human, technical and material resources are therefore concentrated within the production island.

Design and manufacture are organized around these part families so that each product group in manufacturing has a corresponding group in the design section of the factory. The design process is split up according to product families, or their parts, so that the designers perform the whole design process comprising tasks such as determining functional structures and dimensions, geometric modelling and designing new products. Thus, two skill-centred production subgroups will be formed, equipped with local computer assistance and connected by electronic data exchange, namely production and design islands.

The two autonomous subsystems are interlinked by the basic components of CIM architecture. These are:

1. A common database with which all functional programs may interact
2. A data highway to link subsystems
3. Data-exchange interfaces, to integrate subsystems, customer orders and material supply

Unlike the centralized system architectures commonly associated with Taylorist work structuring, these components reflect a CIM environment that is integrated in terms of information rather than control. In other words, instead of formalizing and incorporating almost all production knowledge and work planning into the computer system, the computer system serves as an integrated information system. Although the computer controls routine operations, the planning of work activities is left to the island personnel, who will use their knowledge and skill to optimize island performance with the computer system providing them with accurate information and simulations to support decision-making. The structuring of this information, as it is presented to local workstations and machines, can be changed by island personnel to suit their preferred methods of working and decision-making.

The interlinking of island activities is coordinated by the coordination department, which is responsible for maintaining the flow of information between islands, and for distributing tasks to the relevant design and production islands (as well as initiating new product groups).

The presence of a coordination department does not mean that planning and execution are distinct functions. Orders being produced in a production island do not require detailed process planning information from the coordination department. It is sufficient only to specify the completion date for the part or product and not the sequence of operations. This is left to the members of the relevant production island to decide. Furthermore, there is extensive formal and informal collaboration between all three sectors of the factory by means of electronic data exchange, as well as face-to-face communication.

For example, new product groups will be formed whenever a new product or product family is to be developed. This group, which comprises members from all three sectors of the factory, plans the design and manufacture of the new product. Once the first batch has been successfully produced this group is dissolved. In this way, the product will be adapted to suit the manufacturing capabilities of the production island from the earliest stages of the design process.

On a more day-to-day basis, the interactions between coordination, design and manufacturing enable all personnel to see their work from the perspective of its wider implications for the factory overall, and also to be involved in a wide range of decision-making activities. For example, once product delivery data have been agreed by coordination, design and production personnel, any unforeseen problems arising from within the production island which threaten punctual completion are dealt with by the production island personnel. If the boundary manager for the island detects that rescheduling within the island will not solve the problem he or she may then attempt a form of load sharing with other islands before alerting the coordination department that delivery schedules require reprocessing.

D.2.3 The Production Island

The production island has four main tasks:

1. Production planning (in collaboration with the design group)
2. Production (machining and assembly, plus equipment maintenance)
3. Product and methods development
4. Qualification and training development

The production island work-group comprises a number of "electro-mechanical craftspeople" who will possess skills in machining, assembly, machine setting, NC program generation and editing, inventory management, quality control, machine maintenance and work scheduling and planning. Although all group members will possess these skills, some members will have developed a more specialized interest in certain tasks and functions (e.g. hardware maintenance). Unlike the rigid division of labour associated with the more Taylorist organizational structures, the island has an "organic" division of labour in which individual tasks and skills overlap to allow group problem solving and flexibility in personnel–machine assignment.

Although the group is free to organize and supervise its own behaviour, the internal division of labour follows two guiding principles:

1. The division is always horizontal rather than vertical (e.g. planning and task execution are never divided among group members, each group member being responsible for the setting, programming and machining functions for any given machine).
2. It shall always be possible for every member to experience and develop the relation between quality of product and quality of production (e.g. all personnel participate in the four categories of tasks outlined above).

One person in the group has the responsibility for managing the boundaries of the island. The worker is elected by the group (or the position is filled by job rotation) and negotiates with the coordination department on which jobs to initiate and when they are to be delivered. She or he is also responsible for making sure that the island always has sufficient resources to undertake the work in progress. This leaves the other group members free to concentrate on production (although they will be involved in briefing the boundary manager). The boundary manager does not distribute tasks for others in the group.

When new orders come in, the group as a whole will decide who will do which part of the work and will ensure that every member is aware of their tasks for at least one week in advance to facilitate the planning of additional activities, such as product and methods development and qualification and training development. Typically there will be up to five hours "free time" available each week to allow island personnel to engage in these development activities. These will include brainstorming workshops (to aid product and methods development) and further education workshops (to aid the development of personnel skills and qualifications).

In this process of development activities, care should be taken to ensure that representatives are fully involved. For, as methods like "organization development" are at first sight primarily a workers' participation strategy created by the managment, one should not forget about the different interests of work configuration on the part of the workers (as well as on the part of the management) still being alive.

Also, it is to be considered that the free time concept is in some sense the counterpart of the dynamic productivity concept mentioned earlier. Obviously, it is very important that the higher productivity of the island, as expected by many authors, may by no means lead to increasing worker redundancies. Instead, the concept of participation in organizational development will enable the employees to use their "spare time" for planning not only new methods of higher productivity, but more humane work conditions as well.

D.2.4 *The Workplace*

The worker in the production island group is characterized by his or her craft skills. The worker will typically have responsibility for one machine in addition to access to a common pool of auxiliary equipment and machines. He or she is responsible for designing and manufacturing fixtures and other tools and aids when necessary. In this way, workers will be able to develop their own working methods whilst, at the same time, taking part in the overall development of new products or parts in cooperation with the corresponding design group.

The computer-aided machines which are used within the production island are all operator programmed, although geometric data for parts are typically transmitted directly to the machine database. In no cases are part programs written by the design group.

The human–machine interface has been designed so as to allow the worker to understand the workings of the machine and to develop his or her experience and skill of the machining process. The dialogue between them is not structured by design, but is flexible enough to enable the workers to define their own methods of working. The information system has a network architecture which allows the worker to access all relevant databases and also to receive fast reliable feedback on matters concerning quality, amount of work in progress, throughput times, scheduling alterations and so forth. In addition to the regular meetings and conversations with workers, both within and outside of his group, this information system gives the worker a "panoramic" view of the entire production system. All machines have access to this information system, although some workers may have designed their own personalized information display macros. To avoid confusion, these may only be accessed by the user's password.

D.3 The Scenario Related to the Dimensions and Criteria of Experiencing, Shaping and Evaluating Work

D.3.1 *The Normative Perspective of the Shaping Process*

Every shaping process requires the anticipation of the use-value of products and processes. Moreover, the shaping of the future always creates new conditions and opportunities for the development of individuals in the interrelation between the development of society and of subjectivity. Therefore, it is necessary to establish normative technological and work concepts.

The production concept described in section D.2 is to be understood with this in mind. The global orientation of values in the process of designing HC-CIM is such that the subjective potential for automomous action develops as comprehensively as possible and the ability to (co)shape one's own working and living conditions can be achieved as extensively as possible through computer-assisted work.

D.3.2 *Dimensions of Work*

In the following we propose six dimensions for the structuring of work processes, according to which work can be empirically analysed regarding its relevance to experience and socialization, as well as shaped from the same point of view. A comparison of relevant works from various scientific disciplines shows that there is common agreement concerning the care aspects of work.

Brater designates division of labour, attachment to place of work, the cooperation structure and degree of formalization of the respective working conditions and the extent of insight or participation in strategic decisions as shaping variables (Brater 1984).

Marie Jahoda arrives at five experience perspectives in an evaluation of sociological studies of problems in working life (Jahoda 1983). She designates them "objective experience categories", "as they determine those aspects of work that are experienced by the employed as well as by the unemployed in any manner or form and that have to be experienced whether they want to believe it or not" (see Leithäuser 1986).

These five perspectives are:

1. Imposing of a fixed time structure
2. Expansion of the social horizon
3. Participation in collective goals and efforts
4. Assignment of status and identity
5. Compulsion of regular activity

In their interactive model concerning the relationship between work forms of the personality and personality-related working conditions, Hoff, Lappe and Lempert distinguish between six "work-related aspects" relevant to socialization which they differentiate according to degrees of restriction (Hoff et al. 1982).

All these approaches can be combined into a model, in which six dimensions of industrial life of the work process are distinguished as dimensions of experiencing as well as of shaping and evaluation of computer-assisted work and technology. A formalization of the degree of restriction (such that a differentiation is made between all six dimensions according to standardized, comparable restriction levels) enables one to apply this model to all sorts of work situations. In our case, a qualitative description of six different levels of increasingly restrictive forms of work and production is to serve as a measuring stick for restriction levels. The dimensions of the six work aspects that need to be analysed/shaped are:

1. Time structure
2. Space for movement
3. Social relations
4. Responsibility and control flexibility
5. Qualification
6. Stress control

Time structure includes both time pressure from outside and the degree to which it is possible for the individual and/or group to plan the use of time themselves.

Space for movement includes the degree of explicit formalization of moving from one position to another as a part of the job function. It further includes the implicit possibilities to move if the person feels the need or wish to do so.

Social relations refer to the degree of explicit formalization regarding whom to contact and when, as well as the informal possibilities to communicate across or behind the formal structures.

Responsibility and control flexibility concerns the scope and degree of responsibility placed on the person or group themselves. It includes, as well, the possibilities and actual practice of controlling how this responsibility is managed by the group or individual.

Qualification concerns the functional abilities more or less related to the single job and/or the process of work as a whole. It also includes the more comprehensive aspects of self-renewal and self-transcendence as essential humane abilities.

Stress control includes the degree to which the individual and/or group is able to control the physical and/or psychological pressure that is felt either explicitly or implicitly, in the work organization or man–machine relationship.

Each of the six dimensions must be developed from the perspective of the other dimensions. The relationship between motion and time structure is the first aspect that enables a more precise determination of the quality of the restrictiveness of an existing time structure. In the same way, the social relations, the checking of the work and the perception of responsibiliity are also subject to a time structure. Time structures are also reflected in the stresses and in the learning and overtaxing of abilities. In general, therefore, the question of the time structure is always linked to one of the other dimensions. Analysis and shaping of time structures are thus tied to all other aspects of work (see Fig. 2.9).

Similar preconditions apply to the space of movement. The quality of the possibilities for movement can first be tapped when reference is made to the relationships between:

1. Space for movement and social relations
2. Space for movement and responsibility/control flexibility
3. Space for movement and stress control
4. Space for movement and qualification

An abstractly defined spatiotemporal space of movement would remain relatively meaningless without reference to cooperation, responsibility and work content, as well as to the resulting qualification possibilities. (Please refer to section B.3 for an understanding of the relationship between work and qualification.) In this dimension one must not forget the abilities regarding reflection, oral communication and cooperation which can be classified according to six restriction levels.

Note that the dimension of qualification is better suited than any other dimension as a main indicator for the degree of performance flexibility or restrictivity of a production concept.

The workload results in the same way from the interaction of the other work aspects as well as from the relationship of the subjective provisions (qualification requirements, time structure, possibilities for movement, etc.).

Stress and strain can only be adequately studied and taken into consideration in the shaping process if methods of interpretative sociological research are taken into account beyond the mechanistic concept of ergonomic stress and beyond the traditional concept of empirical, analytical, sociological research.

Each of the work dimensions, classified according to degree of restriction, thus demonstrates a relatively high degree of abstraction regarding the quality of concrete work situations. This is what makes up the analytical quality of the model. In actual research the critical and analytical as well as the shaping-related dimensions of science are becoming more and more frequently understood as interrelated. Thus

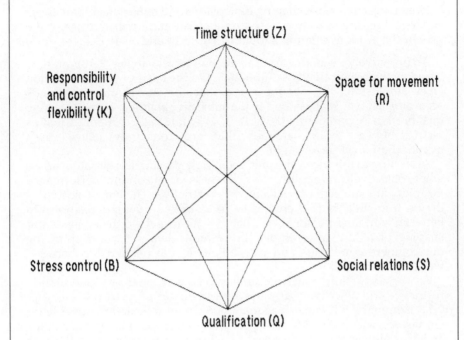

Fig. 2.9. Dimensions of work.

the science of work and industry, for example, consists of the analysis and shaping of work processes and of the interrelation between work and technology.

Regarding the six dimensions of the working-life/work process, one must differentiate:

1. Between the objective and the subjective perspective. The qualification level, for example, can be subjectively described from the point of view of the psychological dispositions, capabilities and development potential of the personality and, on the other hand, from the point of view of the activities constituting the work process (objective aspect). This relationship between work and personality can be applied in the same manner to each of the other dimensions of the working-life/work process.

2. Between the conditional and the shaping (decision) aspect. Each of the six dimensions fundamentally represents a field of conditions as well as a field of possible decisions. In a concrete case, the time structure, for example, can be given and thus becomes a condition for the work and socialization process; in another case with different given conditions, the time structure can represent a shaping dimension. As a result 2×64 (2^6) formulations of a question can formally be distinguished.

3. Between an analytical formulation of the question, such as (1), and a shaping-related formulation of the task, such as (2). The distinction

between objective work/shaping dimensions and subjective discovering/ experiencing dimensions is based on a subtly differentiated use of the grid of the work evaluation shown in Fig. 2.10 and 2.11.

An objectively mean degree of restriction (level 3 in Fig. 2.10) can now be subjectively perceived as "humane". On the basis of the qualification level dimension, this would mean that the qualification demands according to level 3 correspond to the subjective abilities and all excessive qualification requirements (level 0, 1, 2) would be experienced just as restrictively as work demands that are constantly below one's qualification level (levels 4, 5).

According to this view, the only seemingly paradoxical situation arises when objectively minimum restrictiveness or maximum performance flexibility are subjectively perceived as restrictive. The conclusion to be drawn from this must therefore be to assume a dynamic relationship between work and personality. That is, with respect to the analysis and shaping of a human-centred CIM system, always to start from the development conditions and possibilities in the relationship between work and personality.

A work situation which is subjectively perceived as only slightly restrictive and objectively displays a mean degree of restriction (see Fig. 2.11) is humane if it makes adequate performance flexibility possible for employees with very different previous experience and qualifications and if it allows further personal development of employees through expansion of performance flexibility or removal of restrictions.

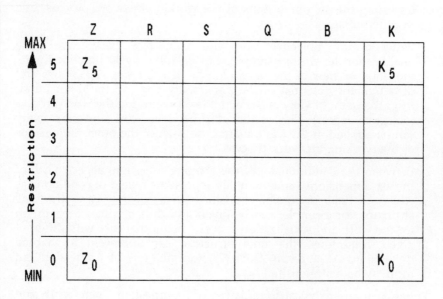

Fig. 2.10. Dimensions for analysis and shaping of industrial work-life.

A wide variety of forms of work and work organization can certainly be found that, if one regards the relationship between work and personality statically, permit "humane" work. However, if one considers the personality to be developing during the work process, then it is of particular importance to create a work organization which enables workers having different qualifications and socialization to take up work, but which, above all, enables the creation of a minimum of objectively and subjectively experienced restrictiveness in the six dimensions of work. The island production concept appears with respect to both aspects (the aspect of heterogeneous initial conditions (e.g. qualifications) as well as the aspect of possibilities for development not only of the working conditions but also of the subjective potential) to be a form of work and work organization deserving special emphasis (Brödner 1985).

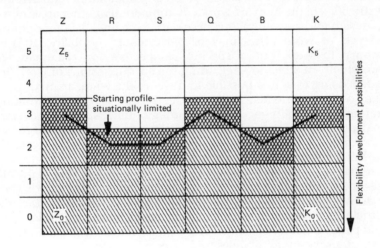

Fig. 2.11. The relation between objective and subjective perspectives on work.

Since the island production concept also proves to be particularly attractive for a broad area of production from the point of view of business management, an interesting conceptual framework is available here for human-centred CIM development.

D.3.3 Example of How to Use the Dimensions of Work in the Shaping and Evaluation Process

On the basis of the present level of socialization research, as well as of industrial psychology and sociology research, one can assume that work situations are subjectively experienced as having the least amount of stress and strain and that the learning opportunities, work motivation and interest in the work are increased if:

1. Work situations are characterized to a certain extent by (solvable) problems
2. Reflection, discussion and experimentation opportunities are provided
3. Understanding and support from fellow men is available
4. Personal responsibility according to one's capabilities is provided
5. These conditions vary from situation to situation and from area of life to area of life (cf. Hoff et al. 1982)

The scenario drawn up earlier enables us to illustrate how to use the six dimensions of work presented in the former section. However, the scenario is, by definition, an ideal type. If the dimensions are to be used in practice both for shaping and evaluation purposes, they must be more differentiated regarding restrictivity versus flexibility. An example of how this differentiation may be developed is shown in Fig. 2.12. This matrix is an example of the classification of work on differentiated levels of restrictions in the humane work environment. For other areas of work (e.g. design and assembly), it is necessary to change the descriptions of the levels of work restrictions or performance flexibility. It is also necessary to work out a specific perspective for a human work environment – a scenario – regarding the six dimensions of work (level 0). After doing this, it is possible to develop the whole matrix.

The scenario presented earlier in this section allows a maximum of performance flexibility regarding spending of time, space of movement, formal and informal communication, self-control planning and carrying out of complex tasks and involvement in decision-making processes concerning stress (level 0 in the matrix).

Alternative scenarios may be drawn up if one prefers one of the other levels or a combination of several levels illustrated in the matrix above. The most restrictive scenario results from level 5. This is characterized by strictly bound and preplanned time rhythm and schedule, strictly bound place, isolation of formally restricted communication, centralized control, sensory–motor abilities and externally steered stress.

Grade of restriction/performance flexibility	Dimensions of discovering/experiencing, and shaping of the work situation						
	Time structure (Z)	Space of movement (R)	Social relations (S)	Responsibility and control flexibility (K)	Qualification level (Q)	Stress control (B)
5 Centralized hierarchical manufacturing concept CAD centred, precision planning objectification of factory work	Time-rhythm bound programme, time schedule	Strictly bound to one place, without possibility of place change, or change in movements	No minimum work related social connections, maximum control (technically facilitated)	Total control via higher factory department; technologically transmitted	Senso-motoric abilities	Stress completely steered externally (technologically transmitted)
4 Operator-oriented manufacturing with a large degree of AV in CAD and work preparation, separate skilled work (machine operators, qualification control, secondary skilled workers)	Not time-rhythm bound, but definite handling times for definite processes	Largely bound to one spot, with markedly reduced freedom of movement, e.g. machine oriented	Formal hierarchical work-related minimum cooperation, no informal contacts possible during work; extensive control (technically transmitted)	Broad control, technologically and socially (personnel) transmitted	Routine fulfilment of job requirements according to a given work programme/rhythm; narrow variation span	Stress largely externally steered (technologically transmitted), little variation in the work speed
3 Classical skilled worker-oriented manufacturing with advanced AV and the separation between machines and secondary skilled workers	Variable time organization within the AV blueprints (e.g. optimizing of CNC programme and varying of technology input data)	Largely bound to one spot (machine group) and occasional shift in space, dependent on cooperation	Cooperation via work-based communication technology, minimal informal cooperation possible; social and technological control	Broad control due to planning structures, and personal control	Controlled, understood and self-optimized work (e.g. skilled machine work)	Stress largely externally controlled (technologically and socially transmitted), variation of the work speed possible within defined borders
2 Classical skilled worker-based manufacturing with limited workshop oriented work planning and centralized material-tool economy, also AV orientation in programming of tool machines	Variable time organization (e.g. in relation to the variation of lots to be handled	Workshop-oriented possibilities for movement (subject to cooperation and work diversity, e.g. by means of secondary skilled work)	Technologically and socially transmitted cooperation on the workshop level necessary informal social cooperation possible to a limited extent (social + technological control)	Limited control (reduced to more complex work situations)	Self-programmed, planned manufacturing on diverse manufacturing installations; experimental abilities (above all, in secondary skilled work)	Work speed externally and self-controlled by work breaks and a combination of work activities (primary and secondary skilled work)
1 Integration of primary and secondary skilled workers. Broad decentralization of manufacturing, with partly autonomous manufacturing islands, largely autonomous workplan, and geometrically based construction (CAD)	Open time organization on workshop level in the framework of definite time sectors (e.g. 14 days) for the lots to be handled, including maintenance, servicing, repair	Good possibilities for movement (necessary because of broad responsibility on the shop-floor level)	Social and technologically induced cooperation and communication with relevant informal components; social control	Largely (cooperative) collective social control	Planning in the frame of part goals within a broader given goal; carrying out of diverse duties/tasks in the workshop	Participation in decisions about stress/burden largely possible; decentralized work planning
0 Completely decentralized workshop-oriented production, with autonomous production islands (partly) integrated construction and work planning	Open time organization of the workshop, and participation in production planning, autonomous workplace oriented work planning	Good possibilities for movement on the shop-floor-level and beyond	Social and technologically induced cooperation and communication (horizontal structure), with well-developed informal components; great degree of self-control	Largely self-responsibility and self-control in a work collective	Planning of complex work relationships; carrying out of complex tasks (in manufacturing and production)	Maximum involvement in the decision-making process concerning stress, possible via decentralized work shaping

← maximum restriction (technocentric) / maximum performance flexibility (anthropocentric) →

Fig. 2.12. Differentiation of the dimensions of discovering/experiencing and shaping of the work situation according to degrees of restriction/performance flexibility, for computer-assisted production work.

D.3.4 Criteria for Human–Machine Interface Design

One of the problems associated with conventional advanced manufacturing systems is that their operation often requires skills that are unrelated to existing skills, with the resultant problems of poor transfer of education and under-utilization of skills.

A human-centred CIM system must aim to utilize existing skills and allow them to develop into new skills. However, skills analysis is an analysis of performance rather than behaviour, a measure of the man–machine system rather than of the human himself in isolation from equipment. Turning skills, for example, have changed over time. The development of the tool rest, the numerical control of servos, and the digital computer have directly influenced the skills involved in lathe operation. Motor skills have largely disappeared with the replacement of hand-wheels by numerical controllers. The problem that must be faced is how one decides which skills to foster and which to degrade.

Galjaard has shown how variety, choice uncertainty and selectivity of information have been gradually reduced in the work experience of the machine tool operator, yet it is precisely these three elements that form the basis of schema development and skill learning (Galjaard 1982).

The aim of a human-centred design is therefore to fit the codes and strategies of the computer processes to the needs and skills of the operator. Our discussion of these needs and skills has been at a very general level thus far, which, at first glance, may appear counter-productive. However, it should be possible to derive general principles of design which can inform specific design choices made in the development of a system. As Eason has pointed out, the degrees of freedom within a design progressively close as the development work progresses (Eason 1982). But this does not necessarily mean that the strategic degrees of freedom available to the operator also close.

An important aim for human-centred design is to ensure that the interface does not constrain the number of useful operating strategies available. Operators should have the freedom to shift strategies without losing software support. The phrase "useful strategies" refers to those tasks where choice uncertainty (i.e. disturbance) is acknowledged to exist and where the tacit knowledge and skills of the operator are used to avert or correct error.

The criteria for human-centred human–machine interface design are divided into three groups, each stressing one of the three following strategies (Corbett 1985):

1. Flexible allocation of functions between human and machine
2. The possibility of a panoramic view of the overall process
3. An open software allowing the worker to follow his own preferred methods of working

1. Flexible allocation of functions between human and machine may be achieved by the criteria of complementarity, operator "control" and interactivity.

Complementarity. Human and machine should help each other to achieve an effect of which each is separately incapable. This principle lies at the basis of human-centred design and stresses that the designer should not produce routine and repetitive jobs for the operator. Such jobs are better suited to the power of computers.

Operator control. This principle primarily stipulates that the operator should have the freedom to choose how those tasks which can be performed by a machine operator, or automatically and/or away from the machine, are divided between the operator and the computer. This allows for a variety of users, a high degree of flexibility, and the opportunity for the user to learn machining skills at a pace consonant with his or her abilities. The flexible allocation approach postulated by Bailey mirrors this principle of operator control (Bailey 1982).

Interactivity. Input and output data should be negotiable and software should therefore allow interaction between the operator (the active data input) and the computer. The level of interaction will depend on whether the task is routine and software-activated (where an interactive screen editor will suffice) or open to disturbance and thus operator-activated (where data manipulation may need to be changed to fit actual, rather than normative, demands).

2. The possibility of a panoramic view of the overall process may be achieved by the criteria of minimum shock, transparency, compatibility and accountability.

Minimum Shock. Common-sense and psychology experimentation alike suggest that error will be reduced if the operator knows what is going to happen next. The principle of minimum shock focuses on system error, when it stipulates that the system should not do anything that the operator finds unexpected in the light of his or her knowledge of the present state of the system.

Transparency. The principle of transparency stipulates that the operator must be able to "see" the internal processes of the computer software in order to facilitate the development of a "schema" (i.e. learning). One can never fully "control" a process without understanding it.

Compatibility. The orientation of modern cognitive learning theory emphasizes that people learn structures rather than isolated pairings (Foss 1968; Carroll and Thomas 1982) and research into command language design emphasizes the need to design system interfaces as coherent structures. Furthermore, evidence powerfully suggests that people develop new schemata by using metaphors to schemata they have already learned (Schneiderman and Mayer 1979; Carroll and Thomas 1982). For the lathe interface it is therefore important that the operator inputs and receives information compatible with his or her training.

Accountability. The operator is entitled to know what the program is doing. To achieve this degree of transparency the software architecture must be self-describing. The integral HELP facility described by Fenchel and Estin is an example of such a system (Fenchel and Estin 1982).

3. An open software allowing the worker to follow his own preferred methods of working may be achieved by the criteria of operating flexibility, disturbance control, fallibility and error reversibility.

Operating Flexibility. The principle of operating flexibility stipulates that the system should offer operators the freedom to trade-off requirements and resource limits by shifting operating strategies without losing software support i.e. the interface should not constrain the number of useful strategies available.

Disturbance Control. Tasks that contain choice uncertainty should be under operator control with software support. Software cannot predict all possible disturbance whereas an operator can cope with the unforeseen.

Fallibility. One of the most valuable elements in the human contribution to a system is the tacit one. This knowledge should not be designed out of the system. Data concerning operations that contain choice uncertainty may be incorrect and therefore the operator should never be put in a position where he helplessly watches the computer carry out an incorrect operation that he had foreseen.

Error Reversibility. In almost all learning environments there will be an element of trial and error, in the exploration of alternatives. Machining errors can be very costly and all but the most highly skilled machinists are, justifiably, not prepared to risk full exploration. Error risk can be minimized in two ways:

1. Strictly define the boundaries of exploration, i.e. restrict the choice of alternatives. An example would be locking out an NC tape editor from the machine and keeping the key away from the operator.
2. Supply information feedforward to inform the operator of the likely consequences of his or her actions. In the context of the other criteria, this option is recommended.

D.3.5 Summary

The efficiency of a human-centred system is based on the complementarity of human and machine. Because of unforeseen disturbances that may enter the system, the operator must be able to control all tasks that contain choice uncertainty via an interactive interface. However, an operator cannot control a system unless he or she comprehends its functioning. A system should support the operator's model of its functioning (schema) so that the knowledge that is needed during infrequent task activity is obtained during general activity.

All routine software-activated functions (e.g. data storage, manipulation and presentation) should therefore be transparent, compatible with existing shop-floor skills and knowledge, self-describing, and predictable. In addition, when the operator intervenes in order to eliminate disturbance he should not lose software support.

D.4 How to Shape and Evaluate Computer-Aided Technology and Work

D.4.1 Typical Shortcomings of Conventional Methods of System Development

The conventional methods of system development presuppose the possibility to develop a description of any work process as a normalized system of functions and data transfers. Furthermore, organizations are perceived to be perfect cooperative systems. When these are formalized it is possible to design a new work process and a new organization based on the technological rationality of "optimal" solutions. The technical aspects of the system are given full consideration, shaped by the mechanistic science ideal.

From a holistic perspective the conventional approach is characterized by too little respect for traditional skills in the domain of both design and manufacture. There is a tendency to consider system knowledge as a superior substitute for, instead of supplement to, traditional skills. This fundamentally narrow outlook causes failure of communication and understanding between system developers and traditional skilled users, which can lead to serious errors.

Furthermore, users of the system have too little influence upon and knowledge of system technology. The reason is partly to be found in the lack of respect mentioned above, and partly the lack of interest and involvement by the users themselves until quite late in the process of implementation. Especially in relation to the HC-CIM approach these shortcomings are to be taken quite seriously, both in a functional and human perspective.

D.4.2 The Search after Methods in Accordance with the New Paradigm

The critique presented above is rather fundamental. It is not just a specific method or technique, but a whole conceptual framework that is questioned. Therefore, it is insufficient only to develop more sophisticated methods, parameters or criteria independent of the fundamental theoretical framework. The development of a new paradigm includes the value orientation, the theoretical concepts and the dimensions and criteria for shaping and evaluating system development. The methods used in the concrete shaping and evaluating processes must be selected in accordance with this holistic paradigm.

One of these methods is the scenario. The scenario is by definition prospective and open for discussion. This method may stimulate both the social and technical scientist to be concrete. Furthermore, the single abstracted aspects are related to a wholeness, i.e. the model of the future factory. It therefore enables a designer to see the kind of work environment into which he must project his end-product. Drawn up as an ideal type, it stresses the opportunities in opposition to the deterministic "one best way" of thinking. Moreover, it may help the social and technical scientists to communicate in a more constructive and creative way because both are forced to make prospective pictures of how their concepts and products may be used and mutually related in the future.

The scenario method is not sufficient. It needs the supplement of analytical and shaping means such as dimensions and criteria of human-centred technology. The six dimensions of work presented in section D.3 are to be considered as guidelines regarding the shaping process of the human-centred CIM system at the organizational level. The criteria for human–machine interface design, presented in section D.3.4, are to be considered as guidelines regarding the shaping process of the human-centred CIM system at the human–machine level.

The guidelines are not fixed in a deterministic way of thinking, but open for discussion. They must nevertheless be reflected both in the shaping and in the evaluating processes of the human-centred system development. By reflection we mean partly a serious consideration of how system development may follow these guidelines and partly an argumentation – in accordance with the general human-centred perspective – if some of the guidelines are not followed.

A human-centred CIM approach includes methods to involve users of the field in the process of system development. As pointed out by Rosenbrock, such methods meet a number of obvious difficulties (Rosenbrock 1983). Workers involved in the R&D process could be placed at a disadvantage through lack of knowledge of these specifications. Also there is the risk that many aspects of the real situation will not be represented in laboratory conditions (e.g. pay systems and realistic pressure of work).

Furthermore, a "Hawthorne effect" may be expected. Moreover, different workers will have different abilities and may need to use the system in different ways. Different management situations may require different ways of using the system, and so on. The above-mentioned difficulties are serious constraints for user involvement in system development. It may be important to reflect very carefully how the interests and practical knowledge of the users in the field are to be communicated to the specialists in order to influence system development.

Firstly, it may be important to plan user involvement as a long-term process, such as continuous involvement over several years, and not limit it to one or two discussion meetings or test procedures every month or two. Second, it may be necessary to use some time, particularly at the early stages of design, to discuss the human-centred idea with the users in the field and have their response to the idea. In this phase of the process the specialists and users have to be quite open regarding different social–psychological barriers which may obstruct the communication. The specialists may learn to understand some of the valuable aspects of the design and craft tradition without romanticizing it.

The users in the field may be open to look at new possibilities of carrying out or organizing the jobs stimulated by methods such as future creating workshops. It is also important to be quite open regarding potential conflicts of interests. Although the human-centred perspective may serve the interests of the user in principle, the concrete suggestions from the specialists may turn out to be a compromise between conventional and human-centred intentions, and thus may run the risk of generating other, unexpected, consequences which may be in opposition to the interests of the users in the field.

Third, it is important to make a flexible plan for the process-oriented approach. The users may have more or less relevant knowledge, ideas, or motivation to participate than expected from the beginning. New problems may appear. If so, it is important to be able to adapt the cooperation between specialists and users during the process as soon as new, more or less unforeseen, problems emerge.

Although user involvement will face difficulties, the consequence is not to avoid user involvement but to find ways to overcome these difficulties. The realization of the human-centred perspective may depend on how well and to what extent the above-mentioned constraints are overcome.

E REFERENCES

Bailey RW (1982) Human performance engineering: a guide for system designers. Prentice-Hall, New Jersey

Braten S (1984) Dialogeus Vilhar: Datasamfundet. Universitetsferlaget, Oslo

Brater M (1984) Forderung der Social und Kommunikations – Fähigkeit im Bereich als Grundlage der Sozialverträglichkeit neuer Technik. In: Hochshultage Berufliche Bildung, Fachtagung 19. University of Bremen

Brater M (1986) Kunstlerische Übungen in der Berufsbildung. In: Projektgruppe Handlungslernen, Reihe Berufliche Bildung 4. Wetzlar

Brödner P (1985) Fabrik 2000 – Alternative Entwicklungspfade in die Zukunft der Fabrik. Sigma Rainer Bohn Verlag, Berlin

Capra F (1982) The turning point. Wildwood House, New York

Carroll JM (1980) Learning, using and designing command paradigms. IBM Research Report RC 8141

Carroll JM and Thomas JC (1982) Metaphor and the cognitive representation of computing systems. IEEE Trans Syst Manage Cybern 12:133–148

Corbett JM (1985) Prospective design of a human-centred CNC lathe. Behav Information Technol 4:201–214

Corbett JM (1987) Human work design criteria and the design process: the devil in the detail. In: Brödner P (ed) Skill based automated manufacturing. Pergamon Press, Oxford

Dreyfus HL (1979) What computers can't do. Harper and Row, New York.

Dreyfus HL, Dreyfus S (1986) Mind over machines. Basil Blackwell, Oxford

Eason KD (1982) The process of introducing new technology. Behav Information Technol 1:197–213

Fenchel RS, Estin G (1982) Self-describing systems using integral help. IEEE Trans Syst Manage Cybern 12:162–167

Foss DJ (1968) Learning and discovery in the acquisition of structured material. J Exp Psychol 77:341–344

Galjaard JH (1982) Science, technology and ethical space. Scitech Research Group Paper, Interuniversity Institute of Management, University of Delft

Habermas J (1970) A theory of communicative distortion. Inquiry 13:360–375

Halfmann J (1986) Die Entstehung der Microelectronik. Zur Produktion des Technischen Fortschritts. Campus, Frankfurt

Hellige HD (1984) Die gesellschaftlichen und historischen Grundlagen der Technikgestaltung als Gegenstand der Ingenieursausbildung. In: Troitsch U, Konig W (eds) Lernen aus der Technikgeschichte. VDI, Dusseldorf

Hoff E, Lappe L, Lempert W (1982) Sozialisationstheorie Überlegungen zur Analyse von Arbeit, Betrieb und Beruf. Soziale Welt 33:29–46

Holzkamp K (1973) Sinnliche Erkenntnis. Historischer Ursprung und Gesellschaftliche Funktion der Wahrnehmung. OPV, Frankfurt

Jahoda M (1983) Wieviel Arbeit Braucht der Mensch? OGB Verlag, Weinheim

Leithäuser Th (1986) Subjektivität im Produktionsprozeß. In: Vollmerg B, Senghaase-Knobloch E, Leithäuser Th (eds) Betriebliche Lebenswelt. Campus, Opladen

Mackenzie D, Wajcman J (1985) The social shaping of techology. Open University Press, Milton Keynes

Moll HH (1979) Zeitgerechte Arbeitsgestaltung. VDI Zeitung 121:1–38

Moll HH (1983) Mehr Produktivität durch weniger Arbeitsteilung. VDI Nachrichten 43:11–29

Polanyi M (1967) The tacit dimension. Anchor Books, New York

Ravden SJ, Clegg CW and Corbett JM (1987) Report on human factors for CIM systems and methods to enhance their usability. ESPRIT Project 534, Technical Report ESP/T87/065/001. Memo 970, Social and Applied Psychology Unit, University of Sheffield

Rosenbrock HH (1981) Human resources and technology. In: Proceedings of the Sixth World Congress of the International Economic Association of Human Resources, Employment and Development, Mexico

Rosenbrock HH (1983) The social and engineering design of a flexible manufacturing system. In: Warman EA (ed) CAPE '83 Part 1. North-Holland, Amsterdam

Schneiderman B, Mayer R (1979) Syntactic/semantic interactions in programmer behaviour: a model and experimental results. Int J Comput Information Sci 8:219–238

Seliger G (1983) Wirtschaftliche Planung automatisierter Fertigungssysteme. Springer-Verlag, Berlin

Simon D (1978) Lernen im Arbeitsprozeß. Der Beitrag von Hacker's Arbeitspsychologie und Piaget's Entwicklungstheorie. Campus, Frankfurt

Weltz F, Lullies V (1983) Menschenbilder der Betriebs organisation. Technik und Gesellschaft. Jahrbuch, Frankfurt

Winograd T, Flores F (1986) Understanding computers and cognition: a new foundation for design. Ablex, Norwood

2.4.3 Reactions of the Engineers to the Shaping Paper

How did the engineers react to the paper? During the spring of 1987, the ISSG held a series of meetings with representatives from each of the national technical groups.

Almost all engineers found the more general theoretical parts of the "Shaping Paper" difficult to comprehend. The following reaction (from a Danish engineer) was typical:

> The "Shaping Paper" presents a lot of new words which I do not know the meaning of, and when I read it several times I had trouble translating it in a way which could help the development of equipment of some kind. I have to translate it into something like a functional specification. This is my way of working when I start up. The gap is too big at the moment.

Some engineers were very critical of the philosophical content of the paper. They feel that concepts such as "life-world" and "praxis" were unnecessary in a discussion of human-centredness. In their view, a justification of human-centred principles and ideas in terms of economic benefits is needed, as "this is the only language that manufacturing organizations (and potential buyers of human-centred technology) understand".

One engineer explained his "feeling" about the paper in the following way:

> It is my feeling that the paper is very much politically related and not so much to the human-centredness of the functions that have to take place in design and manufacturing.

To this the social scientists replied:

> Is it possible to translate such a paper into criteria and not use political orientations? Is it possible to describe human-centredness without a political orientation? That is the question.

There were therefore two kinds of disagreement or confusion between the two disciplines. First, the engineers tend to believe that it should be possible to have a neutral description of human-centredness, distinct from any political orientation, whilst the social scientists considered human-centredness as either implicitly or explicitly politically oriented. Second, in line with their British colleagues, the engineers wanted an operationalized functional specification as a product of the "Shaping Paper", whilst the social scientists suggested the importance of the choice between different kinds of specification as a matter for discussion between all the parties concerned (i.e. engineers, users and social scientists) based on the orientation and scenario of prospective work organization presented in the paper.

Nevertheless, responses to the "Shaping Paper" varied significantly between individual engineers. For example, one of the engineers attempted to apply the six dimensions of work outlined in the paper to the design of the CAM cell operators' tasks. He found this a difficult task and reported that:

> The initial object – to evaluate a list of operator tasks using the social science criteria – was the wrong approach to take. Tasks should not be created and then their "human-centredness" evaluated. Rather, the "human-centred" criteria should be used to generate tasks. Furthermore, the definition of the criteria (and their associated levels) is not rigorous/unambiguous enough to apply in any definitive manner.

In complete contrast, and at the same time as his colleague, another engineer attempted to analyse the design of operator tasks in the CAM cell from a different perspective; namely "with the objective of modelling the operators' workload to establish the number of operators needed to run the cell in a human centred way".

To achieve this, a computer-based model of the use of the operators' time and the effects of work-handling on the lathe was constructed. This model was based on 32 assumptions built into an interpreted Pascal program. These assumptions included: the time taken to load and unload machines, work-handler programming time, unmanned working time, proportion of up-time, time taken for CAM–CAD–CAP liaison, manual finishing times, time needed for training per day, etc.

Results showed that it is slightly better (approx. 2%) to produce an average batch with work-handling than with manual loading. In addition, results indicate that the operator would be occupied for 75% of the shift when loading manually, but for only 50% of the shift when a work-handler is used.

In conclusion, the engineer writes:

> No allowance is made for "non-productive" time such as resting, talking, union meetings and similar activities. A clear conclusion is that with work-handling and negligible non-productive time, the operator would be able to run both machines (each machine takes 218 minutes of his 448 minutes shift time). Allowing for non-productive time and "interference time" during which both machines need his attention, the cell could be run producing only slightly less with one operator than with two.

The report ends with a number of recommendations for using the computer model and program as a design aid. These include a suggestion that "a simulation of a typical task order on a shift could be produced to show what the operator could be doing from minute to minute". Any discussion of the compatibility (or rather incompatibility) between this approach and the work dimensions otulined in the "Shaping Paper" is conspicuous by its absence.

It is clear from these two examples that these two engineers had fundamentally different reactions to the "Shaping Paper". For the first engineer there is a realization that a multidisciplinary design method involves a radical shift away from the conventional engineering design approach epitomized by his colleague's computer model.

2.4.4 The Influence of the Shaping Paper on the Project

Looking back it seems that the scenario or vision of a human-centred factory of the future had the deepest influence on the project. The scenario enabled the engineers to "see" the context of their work.

The CAP group, in particular, overwhelmingly endorsed the passage outlining the scenario. This was because: (1) the group had contributed to the development of vital parts of the scenario, and to the island production concept in particular; (2) the engineers had realized at an early stage that a human-centred CIM concept presupposed an adequate factory model; and (3) there is a higher degree of affinity between CAP and a factory model than is the case with CAM and CAD developments.

But in general, nearly all engineers seemed to accept the ideas outlined in the scenario. Yet they felt that the problem was how to "translate" it into technical, operational terms.

In the subsequent overall system specification, produced in June 1987, the specifications of the CAM, CAD and CAP components correspond directly with the factory scenario outlined in the "Shaping Paper". This report became the point of departure for the second international group established during the project: the International Data Management Group (IDMG).

The role of the IDMG and the experiences of interdisciplinary collaboration in the second half of the project are discussed in the following chapter.

Chapter 3

Interdisciplinary Collaboration in the Second Half of ESPRIT Project 1217 (1199)

3.1 Experiences of Interdisciplinary Collaboration at the National Level

3.1.1 The British CAM Group

As outlined in section 2.2.1, the relationship between the CAM engineers and social scientists was far from ideal in the early phases of the project. However, following the circulation of the "Shaping Paper" and the ensuing discussions, this relationship improved markedly. This was due, in part, to the growing mutual respect that developed between the two disciplines.

For their part, the engineers became aware that technical decision-making was inevitably a slower process when social aspects of design are incorporated into this process. For the social scientists, there was an increased awareness of the need to present these social aspects in a tangible, usable form if their engineering colleagues were to successfully incorporate such qualitative dimensions into the CAM design process.

As a result of this mutual perspective-taking and self-reflection, a design method was developed based on the three main themes of the human–machine interface criteria outlined in the "Shaping Paper". These themes, it will be recalled, were: (1) design for flexible allocation of functions between human and machine; (2) design to enable the user to gain a panoramic view of the work process; and (3) the design of open software to allow the user to follow his or her own methods of working.

Based on the overall system specification (Project report R9), the engineers developed a "framework" for the CNC controllers onto which a "user surface" would be added following collaboration with the prospective users from the user company. A framework differs from a prototype in the sense that the former is the technical (primarily hardware) operating core of the CNC controller which requires the input of machine code in order to function. The user surface refers to the human–machine interface hardware and software. This can take many different forms. The advantage of this design methodology stemmed from the fact that, unlike conventional technical design practices, the framework could be built without closing off many design options in the early stages of design. The design of the CNC lathe software would only be complete after it had been used on the shop floor for a period of time and this completed design may take on a form significantly different from that initially envisaged by the design engineers.

The revised CAM group workplan accordingly scheduled for a minimum of two prototype user surfaces to be designed. The actual number of iterations would depend on the evaluation of prospective users. Provision was also made for the design of a further prototype after the user company had gained experience with the CAM cell over a period of months and made recommendations for changes to the user surface.

Along with their Danish and German colleagues, the UK engineers had uniformly endorsed the scenario or vision of a human-centred CIM system outlined in the "Shaping Paper" as it enabled them to see the context of their work and to visualize how the cell would operate in practice. However, the scenario also appeared to perpetuate the misperception that the CAM cell software was somehow distinct from organizational considerations. In this sense, the scenario was seen as of more relevance to the coordination of work between the CAM, CAD and CAP national groups than to the coordination of work within the CAM group.

Thus, during the first half of the project, the UK engineers focused their attention primarily on ergonomic aspects of the work. It was only when plans for the participation of prospective users were well under way that the engineers were directly confronted with organizational aspects of CAM. This arose because the user company (Rolls Royce, Leavesden) began to raise issues such as training, the role of production and associated line management, and the CAM–CAP interface, once they became involved in prototype evaluation.

In truth, the CAM group work had become so focused on human–machine interface design issues that these important organizational aspects had been left to somehow look after themselves (despite repeated warnings from social scientists). However, once the organizational issues were raised, the CAM group widened its perspective and the implementation process came to be regarded as of equal importance as the technical design process itself. As a result, the relationship between the engineers and the user company took on more importance than the relationship between engineers and social scientists. The role of the latter, during the last year of the project, became one of consultancy and monitoring as a more direct relationship between designer and user developed.

3.1.2 The Danish CAD Group

Influenced by the discussions on the "Shaping Paper" and the user groups, the CAD group decided, in September 1987, to focus on the possibilities of keeping informal communication between the designer and pattern maker intact, whilst at the same time simplifying the means of exchanging ideas and duplicated effort regarding the transfer of the chosen solution to the CAD system. Furthermore, the sketch pad would simplify learning by doing.

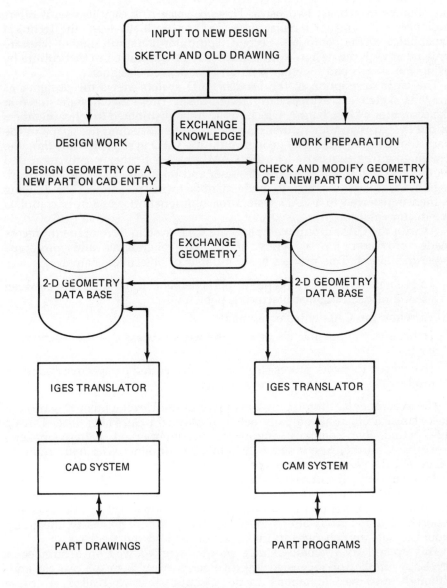

Fig. 3.1. CAD system data flow and structure.

As illustrated in Fig. 3.1, the suggested CAD entry medium makes it possible to use sketches, old drawings and IGES files as inputs from which the design of a new component can be made. The geometry would be calculated and edited either by the designer or the work preparer and exchanged between them. Because the CAD entry system is the same for both, they can easily communicate and change the design in a common language. The CAD entry is portable, thus facilitating informal interpersonal communication.

The electronic sketch-pad-based CAD system also improves learning possibilities in at least two ways. First, because it is easy to use, it offers designers who are not familiar with CAD to gradually learn the technical capabilities of the computer. Second, computer-literate designers who are unfamiliar with the design work culture may be introduced to that culture by using the sketch pad together with an experienced designer.

The basic description of the Danish CAD system views the designer as drawing sketches to express early design ideas. These sketches are stored in the computer's "idea bank", which can then be distributed to other members of the design island for further elaboration. After deciding the shape of the part, the designer draws its geometry on the CAD system. By this time the drawing can be transferred to the CAM area for further examination and modification. In this process the designer and production planner may work together on tolerances, etc., to guide manufacturing personnel. The drawing is then transferred to a CAD system for additional information to complete the documentation.

As with the CAM group workplan, the CAD system development process made provision for a series of design reiterations and prototype experimentation. This process has a number of objectives, namely;

1. To evaluate how the prototype supports personal communication between users and planning/manufacturing personnel
2. To refine the CAD entry equipment
3. To observe the possible changes in the work process and the reactions of users
4. To evaluate the work process from a human–centred perspective regarding productivity, learning and work quality.

The experimental prototyping took place in the Development Department in a Danish industrial company between May and December 1988. During this period five company employees (designers and pattern makers) participated in the experiment – including the designer who had originally suggested the sketch pad concept.

Before the experiment began, the user groups and the CAD research group discussed how to structure the process. It was agreed that the designers should try to use the sketch pad in carrying out their normal daily work tasks and report back to the group after a couple of weeks. Where necessary, the group would initiate any changes to the system based on the feedback.

This cyclical try–evaluate–change process worked well on a number of levels. A number of suggested improvements arose from the process and, although resource limitations made it impossible to realize all these suggestions, a number of important modifications were made. The "spin-off" effects of the process were equally important. During the evaluation of the sketch pad, a creative and very stimulating atmosphere developed in the group (two designers, one engineer, and two social scientists). A number of ideas on how to improve work organization and educational opportunities were discussed during the six-month prototyping period. These were not foreseen or anticipated, but were extremely informative as they led to a continuation of discussion into the spring of 1989.

These discussions elaborated some visions of the "ideal" design office which had arisen during prototyping of the sketch pad, and inspired both engineers and social scientists within the CAD group. As a result a series of meetings were held which contributed to an understanding of theoretical knowledge, "knowledge through experience" and practical knowledge.

Thus, the CAD group experienced a gradual development of a positive shaping atmosphere during the third year of the ESPRIT project. Beginning with an atmosphere of distrust, prejudice and rather uncoordinated activity amongst the engineers and social scientists (provoking a crisis in the spring of 1988), the "Shaping Paper" and the influence of the user groups had a profound effect and helped establish a more or less continuous dialogue and interdisciplinary collaboration. This contributed to a more holistic paradigm for the development of human-centred technology.

3.1.3 The German CAP Group

The actual development of the shop-floor monitoring and control (SMC) workstation was carried out without truly interdisciplinary collaboration between engineers and social scientists. This was not so much due to explicit objections to interdisciplinary research and development, as to patterns and habits characteristic of research and development which are confined to individual disciplines. These were not sufficiently reflected upon with regard to their limiting effects on collaboration. This lack of reflection reinforced two related elements of traditional development practice which are directed against "integrated development concepts".

First, from the engineering viewpoint, only the more or less "finished" product or component represents a "real" contribution to collaboration worthy of discussion. The product facilitates communication, it is self-explanatory in the sense that it is obvious what one is talking about. Talking about "unfinished products" – which may be translated as meaning something beyond the objectivity of technology – was typically experienced by the CAP engineers as limiting their own professional strength: "Let's first realize a finished version before representing it for discussion and criticism".

Secondly, the fact that the SMC prototype which was developed by the academic partners during the second half of the project showed noticeable traits of "human-centredness" was chiefly attributable to the comparatively intense discussions, both formal and informal, which took place at the international level (e.g. the International Data Management Group, the International Social Science Group, project meetings and international workshops). This continuous dialogue about human-centred CIM, which was quite divorced from the actual development and design work of the CAP group, eventually reinforced a view of the problem and the problem-solving horizon which carried over into the university-based designers' activities. Indeed, the lack of interdisciplinary collaboration within the CAP group was partly compensated by the influence of the discussions and presentations at the international level.

It should also be noted that the hierarchical nature of the research and development process, which hampered interdisciplinary collaboration during the first half of the project, was not fully resolved during the second half (section 2.1.3). Paradoxically, the absence of professors from project events may also lead to innovative phases in the research and development process.

This absence was accompanied by an intensification of social relations within the CAP group, which ultimately led to a certain innovatory base for interdisciplinary cooperation. It was not least to this effect that project participants attributed the creative developments and refinements of the SMC workstation concept which were achieved during the second half of the project.

3.2 Experiences of Collaboration at the International Level

Whilst the collaboration between social scientists and engineers showed marked improvements during the second half of the project, it is instructive to examine the experiences of collaboration between the national groups.

In section 2.4.1 the differences between the approaches of the three national groups were discussed and reference was made to the importance of practical shop-floor knowledge for the work of the CAD and CAM partners. The emphasis of the CAP group, on the other hand, was more firmly rooted at the organizational level of analysis.

For example, the main emphasis of the CAM work was the establishment of a fully operational CAM turning cell comprising two CNC lathes, automatic parts-handling and tool-changing equipment and a cell controller workstation. This cell was viewed as a basic building block for the creation of production islands.

Compared to the CAP work, this emphasis on technical design and configuration is rather narrow as it fails to explicitly incorporate a consideration of wider issues such as organizational design and systems architecture.

These two approaches are clearly complementary but the precise nature of the relationship bewleen organizational and technological design was rarely touched on until the CAP argued the case for a human-centred CIM demonstration site at Bremen which would contain a variety of conventional NC and CNC machine tools at a board meeting. This proposal provoked intense criticism from members of the CAM group who felt that the utilization of technique-oriented as opposed to human-centred machines in such a demonstration would undermine the very nature of the project's human-centred CIM philosophy.

The CAM group regarded the production island concept (and its elaboration by the CAP group) as a crucial organizational context within which to place human-centred technology (of which the CAD group's electronic sketch pad was heralded as an examplar).

This difference in approach was to have more than a purely academic significance. Consider the shaping criteria and scenario outlined in section 2.4.2. These effectively ask engineering designers to leave design options open and to promote a maximum degree of flexibility in the use of the technology they are designing. When it comes to designing a machine tool interface, users can become involved in answering the question "which design options should be left open and for how long?" However, when it comes to designing a complex CIM system, user participation becomes more problematic owing to the technical complexities and organizational considerations which must be woven into the design specification and process. Yet the placement of human-centred technology within a centralized, rigid systems architecture may result in a loss of the former's intrinsic flexibility and tool character.

The "Shaping Paper" offers only general guidance on the issue of systems architecture design and it was whilst in the process of writing this book that the authors came to realize that this was a fundamental oversight which contributed, at least in part, to the CAM–CAP group conflict discussed above.

Subsequent discussions with engineering colleagues revealed that the lack of a specific human-centred CIM architecture (or even a detailed technical scenario) was perceived as a problem which jeopardized the development of a truly integrated human-centred system specification – a key project objective. It is perhaps of historical significance that in the original (rejected) project proposal (section 2.1) explicit financial and strategic provision had been made for the participation of Intervisie (a Dutch consultancy firm) in tackling what the proposal termed the "organizational interface" issues. Unfortunately, Intervisie withdrew from the project following the rejection of the first proposal by the ESPRIT directorate in late 1984.

With the benefit of hindsight, the differences between the CAD and the CAM groups on the one hand, and the CAP group on the other, may have been resolved far earlier in the project if the common ground between the two perspectives had been explored. This common ground – where technical and organizational facets of human-centred CIM come together – is systems architecture, and it was through the work of the International Data Management Group and the use of "shaping workshops" (Chapter 4) that this ground was established.

3.2.1 The International Data Management Group

The IDMG was established in July 1987 as a subgroup of the national technical groups. Its objective was to elaborate "a reference model for human-centred CIM data management" and thereby to achieve consistency between the software and hardware developments within each of the three national groups.

From the outset, the IDMG comprised only engineers who were inspired by some reference models developed by the US Air Force in the mid 1970s: integrated computer-aided manufacturing (ICAM) and structural analysis and design technique systems (SADT). The basic outline of the IDMG reference model is shown as Fig. 3.2. This approach was not generally approved of by the project board and others as it was argued that the implicit factory architecture of the approach was in contradiction with the "Shaping Paper" scenario.

After discussions within the IDMG, the engineers specifically asked for social scientists to join the group. This occurred in October 1987 when two social scientists joined the group (one of whom was later to become chairperson of the group). However, this did not resolve the underlying problems and, in January 1988, the IDMG was formally requested to incorporate key elements of the overall systems specification into their work.

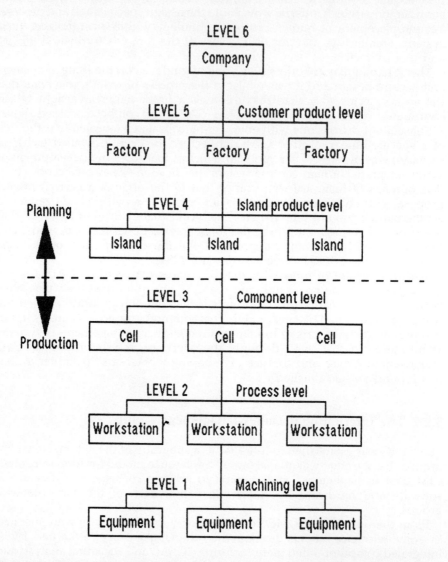

Fig. 3.2. International Data Management Group reference model for CIM data management.

Thus the IDMG had the difficult task of developing a data management while at the same time referring to the concept of human-centred CIM.

In the autumn of 1988, the IDMG produced a report entitled "A reference model for human-centred CIM data management". In the report it was stated that "some of the SADT and IDEF (ICAM definition) tools are used, but with an aim truly unlike that of the ICAM project". (It should be noted that the basic assumption of the ICAM project was that every act of manufacture and managerial control can be represented by data.)

This report documents how the IDMG managed to transform general qualitative criteria and orientations into a data management model. In the introduction to the report, the group critically analyse the objectives and methods of the ICAM project and clearly state that, despite their rejection of the ICAM approach, they have adopted some of the methods and tools associated with it as a basic starting point. This difference in objectives is made clear in the introduction to the IDMG report:

Computer technology itself does not determine any specific shape of manufacturing. Yet, for the development and use of computer-aided human-centred integrated manufacturing systems, a constant rethinking of system development, philosophy and methods is essential, as is experimentation with different forms of work organization and planning.

Thus, instead of the ICAM philosophy, the IDMG refer to ESPRIT project documents:

The human-centred CIM project is based on the theory of experimentation with group technology and autonomous island production. However, CIM transcends the boundaries of the production island and emphasizes the integration of design, planning and production. This makes experimentation with organization and system development much more difficult because a century of systematic division of labour within the factory has left us with little knowledge of, and experience with, the integration of the technical office and the shop-floor.

The IDMG concludes that:

The steady evolving possibility of technical integration of computer systems in the different factory sections compels the necessity of a new understanding of all aspects of manufacturing. Without this understanding, technical integration will be either useless and therefore totally uneconomic or its consequences unpredictable.

Information technology is thus considered as the vehicle for the extensive reintegration of planning and production. But the success of this reintegration will also depend on "development of another kind of collective skill effecting design work, planning work and shopfloor production work. Understanding collective work and collective skills is essential for the development of human-centred CIM systems".

In common with the "Shaping Paper", the IDMG stress a process-oriented development towards integration. Thus it is presupposed that the cooperation between different sectors of CIM work will gradually undergo a number of changes and that integration of different skills may become possible; "perhaps ending up with the integrated education in which subjects are not design or manufacturing, theoretical or manual".

A similar understanding is expressed in the view that a CIM project must facilitate the learning-by-doing of the beginner and let the skilled user "define the principles according to which the code be made and then make it, only asking for additional information when needed".

The IDMG are aware of the far-reaching implications of this and argue that the biggest obstacles to be surmounted are the traditional division between white- and blue-collar work and conventional managment wisdom which equates efficiency and performance with job simplification.

All the work, decision and learning processes which are intended to find a place in the island production groups are described in more detail than in the general factory scenario outlined in the Shaping Paper. Moreover, this description is structured so as to facilitate the integration of human activities with data structure and information flow.

However, this codification of production processes into the open structure of the data management model enables a different kind of implementation, namely increased control over the production process as well as a restriction of opportunities for learning. This approach, which runs counter to the tenets of human-centredness, is seen as one possible projection of the concept of data management into "work-life".

It is precisely in order to counteract this possible misinterpretation that the IDMG drew up a detailed scenario outlining all their specific assumptions: "To foresee the consequences of new technology and work organization, a detailed scenario has to be written and discussed at each special implementation. Especially when new software is developed, the scenario should be very detailed as it, together with the flow diagrams, can tell how to make the data model, the systems orders and interface."

Thus the detailed scenario is perceived to be both an orientation for the technical development of the data management model and a means to evaluate the implementation process. If so, it may be seen as a new way of developing and implementing CIM systems.

Certainly, there are indications that the authors of the IDMG report conceive of the detailed scenario as a necessary element of a reference model rather than as one possible projection of the data management concept into work-life. If this is so, the model proposed represents a fundamentally new form of describing and developing CIM systems. The data management concept alone is no guarantee that the factory scenario outlined in the report will be realized. The scenario itself forms the core of the factory model. This core provides important hints about a company's decision-making and management structures, the content and forms of in-house enskilling and training and the drawing up of industrial relations codes of practice. What it does not offer, however, is any guidance on how the design and implementation process may be carried out in practice. In this sense the IDMG report has more relevance to the development of a "green field" industrial site than to a company who already have some basic CIM building blocks and entrenched organizational practices. This issue is discussed in more depth in Chapter 4.

Overall, the degree of human-centredness of the IDMG approach depends on the extent to which it successfully encourages a participative process involving the end users' imagination of prospective paradigms of work, tools, communication and opportunities for learning. Perhaps more fundamentally, it is important that technically mediated integration per se does not have a distinct ontological status, but may be questioned if alternative, more human-centred, solutions are available (Chapter 4).

To conclude, the Reference Model for Human-Centred CIM Data Management report can be perceived as a systematic response to the concepts and scenario presented in the "Shaping Paper". In this way, the dialogue between the social science and the technical groups was realized in the ESPRIT project at the international level.

3.3 Lessons Learned

3.3.1 Constraints on Interdisciplinary Collaboration

It is clear from the preceding sections that the engineers and social scientists within the project experienced varying degrees of difficulties in working together. This is not surprising given the difficulties experienced by ergonomists (who typically focus on comparatively narrow aspects of work-life such as health, safety and performance) who work with engineers (see Corbett 1990).

Almost without exception in the early stages of the project, the engineers in Project 1217 (1199) viewed social science and/or user input to the project as something which could be "slotted" into conventional engineering design practice. In other words, their expectations centred around the development and use of social science and/or human factors techniques such as design criteria. This explains why the third section of the Shaping Paper (which details the shaping dimensions and human–machine interface design criteria) was accepted far more readily than the earlier, more philosophical, sections.

To understand why this should be the case it is worthwhile to return to a consideration of the philosophical, ideological and cultural roots of engineering design discussed in the Shaping Paper, in the light of the ESPRIT project experiences.

At least three main areas of (philosophical) incompatibility between engineering and social science approaches to design may be identified from the preceding sections of this chapter. These focus on differences in logic, epistemology and assumptions about human nature.

Logic

Perrow argues that an analysis of the decisions of designers, partly determined by their supporters but also instilled in them in their professional training, reveals that they work from a design logic rather than an operating logic (Perrow 1983).

The design logic employed by engineers and the operating logic employed by users and social scientists are to some degree contradictory.

> A good design is compact, but good operating logic stresses easy access to controls and to system-state information . . . A good design favours information on subsystem dynamics, because it is easier and cheaper than total system dynamics and does not require integration with subsystems someone else has worked on, but good operating logic emphasizes information on the dynamics of the total system. (Perrow 1983, p.528).

The CAM group engineers' use of the criterion of "consistency" is an example of this design logic. Consistency is a functional characteristic of a human–computer interface which produces a "clean design" in terms of design logic, but which standardizes the activity of the user and bears no relationship to user knowledge, experience or skill. Certainly the idea of open software in which users are free to define their own macros and work sequences (discussed in Chapter 2) is not readily seen as clean design by engineering designers at the present time although there are signs that this is changing within the computer science community (Ehn 1988).

At a different level of analysis, the experience of the CAP group illustrates that industrial and academic design engineers may work from different logical standpoints. It will be recalled that the "project dualism" which resulted in the development of two SMC workstations arose primarily through the industrial partners' emphasis on the logic of the marketplace.

Epistemology

A further incompatibility between social science and engineering approaches to design stems from the type of data and knowledge each discipline employs. Social science deals with considerable amounts of qualitative data, whereas engineering design data is more quantitative.

This difference was most apparent during interdisciplinary discussions on the "Shaping Paper." It is part of engineers' self-image to be concerned with maximally precise facts and, in the case of shaping, to be dealing with criteria that can be and have been operationalized in a quantitative form. Whilst the engineers were generally appreciative of the factory scenario outlined in the paper as a backdrop reflection, they effectively ignored the shaping dimensions in their design work. At first sight this would appear to be contradictory as the scenario is of a more qualitative nature than the shaping dimensions. However, it became clear that the scenario was primarily viewed as a description of the overall design objective. As such it impinged very little on their design practice at a detailed level. The dimensions, on the other hand, were offered as aids to design decision-making. Their qualitative nature, however, rendered them unusable from the perspective of conventional design methodology. It is significant that the (quantitively derived) computer-based model of CAM cell operation developed by one of the CAM engineers was far more readily accepted by his colleagues as a design aid than the shaping dimensions.

The engineers' reactions to the "Shaping Paper" reveal the more general difference of orientation between the social and engineering sciences. The epistemological tradition of engineering science distinguishes between problem areas suitable for objective, quantitative analysis and areas which contain values, attitudes and feelings. The abstract mathematical form of thinking reinforces the propensity to objectify, quantify and abstract from the social science paradigm of values and social interests.

As described by Rosenbrock:

On the technical and economic aspects the engineer can let his imagination range freely among the options, and a good combination of these options will often present itself to him . . . By contrast if he tries to include the social aspects of the options open to him, he has no general principles which he can use for guidance. He is explicitly warned against reliance upon values which might be thought to serve no place in general principles. (Rosenbrock 1983).

This was exactly how the social scientists experienced the engineers' ambiguity towards the "Shaping Paper". It was clear that the engineers did not know how to cope with the paper. Some of them tried to quantify the different dimensions and human–machine design criteria individually but gave up because they found that "you have to take the whole unit". When unable to isolate and quantify the dimensions, they were also unable to establish a measurement yardstick with which to evaluate design solutions.

At a more macro level, the engineers were not convinced that social science concepts and objectives could be realized as economically viable strategies. In some way the economic paradigm of competition (e.g. "automate or liquidate") seems to have similar characteristics of universality to the natural sciences. If this is accepted, the engineers' natural science ideal of objectivity is intact and it is therefore possible for the engineers to legitimate their "scientific" orientation: to develop and implement technology in accordance with the "universal economic laws" of industrialized society.

Assumptions about Human Nature

A third area of potential conflict between engineering and social science design stems from different implicit or explicit assumptions about the behaviour and needs of human operators.

In conventional engineering design philosophy, human behaviour and needs are not explicitly considered. The drive to achieve system robustness and performance predictability and controllability results in operating tasks which themselves are predictable, standardized and regulated by prescribed rules and procedures. For many designers the human is the major source of unpredictability and error in systems operation – hence the logic of automation.

Even though the engineers in the project generally did not express this view, in many instances the incompatibility between the social science view of human nature and work-life and that implicitly held by the engineers became visible.

For example, from the onset of the project, it was clear that the engineers and the social scientists shared a very different understanding of the nature and role of communication. For the engineers communication tended to mean the transfer of technical data and information between workstations (data management), whereas for the social scientists communication referred more to interpersonal, primarily verbal intercourse between social actors based at these workstations and elsewhere. These two models of communication reflect images of the human as a mechanistic information processor and as a formulator of plans and strategies within a shared world of symbols and language, respectively.

However, it was agreed that both forms of communication were of great importance for efficient system functioning, although the engineers felt that too strong a reliance on informal communication networks within a system would make that system prone to unreliability, unpredictability and conflicting sets of instruction. For the social scientists, on the other hand, an over-reliance on electronic data exchange as the medium of communication within a system would make the overall system uncreative, unresponsive or insensitive to non-informative information, and reduce the richness and quality of work-life for users.

In discussions about the importance of enhancing, rather than diminishing, the role of face-to-face communication in the CAM cell, for example, it became clear that many of the CAM group engineers felt that there was little benefit, in terms of performance, to be gained from such a strategy. This view led them to accord a low priority to such issues. Indeed, this low priority to the role of interpersonal, compared to human–computer and computer–computer communication was also reflected in the documents produced by the International Data Management Group.

Discussions about communication within the CAD group revealed a similar antipathy between the engineers and social scientists, and it was only after the development of the electronic sketch pad concept that the two disciplines developed a common ground. Significantly, it was the impetus created by this tangible design solution which facilitated this working consensus rather than the philosophical justifications presented in the "Shaping Paper".

3.3.2 Experiences of Improved Collaboration During the Project

All three national groups experienced a crisis in their collaborative endeavours during the first year of the project, partly due to the constraints discussed in the previous section. With the exception of the German CAP group (where a project dualism developed and the academic and industrial partners effectively ceased collaboration), the crisis was the point of departure for profound changes in the development strategy of the technical groups. Furthermore, the collaboration considerably improved during the second half of the project. This does not mean that the constraints disappeared. But the common understanding, sympathy and will to collaborate was exercised when agreements were reached regarding the functional perspective of the technically led approach.

Hence, when the sketch pad idea was agreed upon within the Danish group and agreements were made with a Danish company to test and develop the concept, the aforementioned constraints suddenly became inspiring topics for discussion. Both parties knew that they had different experiences and ways of expressing themselves, but in the continuous dialogue with the company's designers these differences were not incompatible but complemented each other.

It would appear from the experience of the Danish and British groups that successful collaboration depends on at least three criteria being fulfilled. First, all group members should have a commitment to the human–centred approach. This is fundamental and, in the case of ESPRIT Project 1217 (1199), despite a few disagreements during the early months of the project, this criterion was fulfilled.

A second important criterion for successful collaboration is that engineers and social scientists share a common vision of what the resultant CIM system should look like. The scenario outlined in the "Shaping Paper" consolidated this shared vision and was to prove crucial at both the national and international level.

The third criterion is that project participants should share knowledge and understanding with regard to the practical steps that need to be taken to achieve collaboration. This criterion relates to methodology and failure to fulfil this criterion during the early stages of the project proved to be the biggest constraint to successful collaboration. In the case of both the British and Danish groups it took an "exemplar" to overcome this constraint. Thus the "open CNC framework" concept in Britain and the electronic sketch pad concept in

Denmark proved to be crucial turning points in the collaborative process. It is interesting, and perhaps fundamental, that in both cases user participation was the key methodological breakthrough. Users tended to be relatively free of the philosophical and methodological biases which hampered the engineering and social science project participants.

At the international level, the response and elaboration of the ideas presented in the International Social Science Group's "Shaping Paper" and the International Data Management Group's "Reference Model" report demonstrate that constraints on collaboration can be overcome and serve to stimulate the dialogue between engineers and social scientists.

Looking back at the process over the project's three-year life-span, it is possible to see a number of prejudices and unreflected decisions and activities which resulted in unnecessary tensions and wasting of resources. These have been discussed in some detail in this chapter. However, as a pioneering interdisciplinary approach with a human-centred orientation, the process created a number of negative and positive experiences, some of which inspired the authors to analyse how prospective interdisciplinary design may reflect and utilize different kinds of attitudes, knowledge and experiences in an open and transcendent dialogue process. The following chapter examines this in more detail.

Chapter 4

Crossing the Border

In this final chapter, we attempt to describe the concept of "crossing the border" in such a way as to render it transferable to other shaping projects. This requires, on the one hand, abstracting from the specific experiences gained in the project on 'human-centred CIM systems", while, on the other, concretizing the paradigm of computer-aided participation (Rauner 1988a). Our concept is not simply an idealist notion that we use in order to oppose traditional factory development and the interests guiding it. Rather, this concept refers to a realistic option in a historical branching-out situation whose feasibility is supported by a number of indications.

After forty years of comparatively steady and obvious growth of our production system, we are apparently on the brink of a new development stage, which, in Lutz's words, resembles a giant experimental terrain (Lutz 1988). None of the factors involved in factory development, neither production technology nor computer-aided production work, nor the level of available qualifications indicates the direction in which the "new" factory will develop. Any attempt to extrapolate possible future production systems from one of these factors has been futile.

The terms deregulation, flexibilization, availability, as well as open system architecture, mark the evaporation of continuity and clear-cut development paths. Instead, these terms refer to changes that have to be coped with.

The concept of "manufacturing consent" (Burawoy 1979) encourages us to present, and to invite the public to discuss, our understanding of "crossing the border". We use this phrase to denote the self-critical extrapolation, from our experiences in the HC-CIM project, of a new development tradition of computer-aided participation.

4.1 Crossing the Border – a Tentative Definition

4.1.1 Fields of Action

In our project, the founding of a new development tradition occurred rather laboriously through interdisciplinary cooperation and participation. The process was aided, however, by the fundamental experiences and reflections published by Ehn (1988) as well as Winograd and Flores (1986).

Crossing the border does not simply denote a particular way of participatory systems development and software engineering. Computer-integrated manufacturing challenges everybody involved to shape the integration of "new" and "old" technologies and to try and achieve a well-poised balance in the complex interrelation of the development of technology, work, work organization and training of staff. This well-poised balance corresponds to a social modernization strategy in which it is the interests of the people concerned that count and that determine development objectives rather than vice versa. Applied to factory development, such a strategy relies on readiness to innovate on the part of the people concerned. Inspired by intellectual curiosity and keen on experimenting, they actively grapple with what is technologically possible.

Below, we sketch the dimensions implied in the concept of crossing the border.

Leaving aside the shaping object – here, computer-integrated manufacturing in the sense of a complex computer-aided work system – there are two dimensions which delineate the area in which conventional limits of thinking and acting have to be transcended as a precondition for shaping in a human fashion the computer technology and computer-aided integration that are increasingly coming into use in the factory.

	shaping competence agents	attitudes (interest, motivation needs, orientations)	abilities (knowledge, experience, skill)	actions (shaping-oriented communication, co-operation, intervention)
society	social and government institutions and their representatives			
	external people concerned			
companies	users (wage-earners)			
	employers (management)			
academia	engineers and engineering scientists			
	educationalists, doctors, scholars concerned with humans			
	sociologists, social scientists			

Fig. 4.1. The qualifications enabling agents to transgress limits in the course of shaping.

The areas thus defined represent, first of all, the questions raised by the elaboration of the concept of crossing the border: which attitudes and abilities do social scientists, engineers or users require in order to be able to collaborate – on the shaping of humane computer-aided work systems – with the other agents by contributing their specific, work-related interests, needs and abilities to an integrated and participatory development process? Any answer to this question presupposes a clear definition of the understanding the people involved need to have for each other in order to gear the accumulation of experience and the adjustment of interests towards the given development objectives.

4.1.2 Agents Involved in the Shaping Process

The agents involved in participatory and integrated technology development represent seven groups of people from three social sectors: state and society, companies, and academia (see Fig. 4.1).* We shall leave aside the agents from government and social institutions, despite the fact that their attitudes, abilities and activities are equally relevant to crossing the border in participatory and integrated development. Here belong the ESPRIT programme and its representatives, for instance. One would have to analyse how the intentions of the programme as well as budgeting and project processing rules bear upon agents' willingness to award funding to a given R&D project. Here also belong trade unions and employers' associations, whose agents have a great impact on any participatory and integrated (in the sense of transgressing limits) shaping of work and technology.†

In the HC-CIM project, the people involved in shaping came mainly from companies and academia. The experiences gained in this project confirm the validity of our approach. This consists of involving in the shaping of computer-aided work systems scholars concerned with humans, social and engineering scientists as well as (potential) users. The object of the shaping process – a human-centred CIM system – cannot be reduced to an instrument of merely technical dimension; it is also of considerable social relevance. This is why the project started on the assumption that the involvement of social scientists and of scholars concerned with humans would be an advantage. The usefulness of interdisciplinary development groups of projects appeared to us to be theoretically founded on our understanding of "work and technology", as explained in Chapter 2. Empirically, the value of such groups had become evident in other projects. Kling and Scacchi (1982), for instance, talk about a "web of computing" with regard to the development of computer-aided work systems. This phrase denotes that:

the specific network of people, resources, services and organizational structures which mediates computer use, is an inherent component of the system itself. Computer-based technologies cannot be assessed apart from the cultures and organizations that use them; they are best considered therefore as deeply embedded in a network of organizational and social relations,

---

*See also the areas of technology shaping.
†We mention here only in passing the fact that the task force came round to accepting a project with an emphasis on "socially compatible technology shaping". This transgression of programme limits goes to show the potential flexibility of state agents, rather underdeveloped though it commonly is. The rigid demarcations of departments and programmes usually stand in the way of innovation.

in a framework of activities and institutional arrangements. In this final configuration, computer-based technology is a form of social organization. (Gordon and Kimball 1985, pp 79–80)

This view concords with our understanding of technology and confirms the abstract recognition that technology shaping is the "concern" of various academic disciplines. The question as yet largely unanswered, however, is how the competence of the highly segregated academic disciplines, or of their representatives, has to be enlarged in order to be of use to a shaping of work and technology that is both participatory and integrated. Phrased more dynamically, the question is what learning processes need to be involved in shaping.

What can and must "crossing the border" mean to the agents involved? This question will be exemplified below with regard to engineering and social scientists as well as users.

Two of the boundaries that engineers have to transgress in order to use their shaping competence more extensively than has hitherto been the case will be considered more closely:

1. Engineers' "technico-scientific problem-solving horizon" (Hellige 1984), blindly accepted by the individual, as well as its evolution as one of the dimensions inherent in industrial–cultural development.

2. The reduction of technological science – which always implies a specific concept of technology – to the logic inherent in what is technical.

The example of engineers' technico-scientific problem-solving horizon shows clearly how engineers and engineering scientists have to transcent the limits of traditional problem solving. The term introduced by Hellige points in the right direction:

I use this term to denote the consensus that exists in the individual disciplines and schools within the engineering community. This consensus changes over time and is checked by economic, social and political factors, professional traditions, effects of base technologyies and the "scientific styles" of problem-solving behaviour. The "technico-scientific problem-solving horizon" thus represents a historically and socially determined limitation of the sum total of the knowledge and experience available to any one branch of technology. Consequently, only a limited number of the total amount of feasible technical solutions, objects, and methods are seen as possible and hence elaborated. Depending on the given problem-solving horizon,

certain development paths are excluded right from the start of a given assignment and problem definition;

economizing on material and energy costs takes precedence over work-saving technological progress in the development process;

certain building and system sizes are seen as standard norms and monumentalization predominates over minimalization or vice versa;

centralized, hierarchical, hierarchically distributed or rather more decentralized and autonomous solutions are given preference in energy, information and work systems;

the individual stages of the technological building process, the scope of alternative solutions considered and the predominant criteria of assessment, according to which optimization and selection occur, vary;

through the actual design of the technical structures of hard- and software, it finally determines, to a considerable extent, the possible scope of use and the chances for shaping the given human/machine relations differently. (Hellige 1984, p.10).

In this context, crossing the border involves becoming aware of one's own and of the culturally predominant technico-scientific problem-solving horizon, as well as the concomitant restriction of shaping perspectives. The re-evaluation proposed by Hellige affords the opportunity to understand and assess the technological alternatives selected – as opposed to those ignored – as expressions of objectified interests, purposes and problem-solving horizons, and that beyond the study of the historical evolution of any technical artefact.

This approach already points to the transgression of the second restriction, i.e. the understanding of any technology as a means–end relation. As a functional relation, technology always also bears a use-value. The shaping of technology is concerned with concrete technical artefacts and thus with the realization of use-values. This requires agents to become aware that technology is tied to specific interests and purposes and that it is moulded by a given culture. Moreover, the shaping process has to be geared towards usefulness and functionally as the fundamental requirements of any technology.

In traditional software engineering, it is at the design stage that social criteria are translated into technical ones (see Fig. 4.2). Designing presupposes descriptions of the qualities required of a given system – its use-value qualities – and of the criteria for testing them. Designing is essentially an act of transforming the use-value qualities desired into a technical description of the system to be realized. In the case of software engineering, this transformation involves defining the systems structure, interfaces, data and control flows, and the division of components.

This transformation of social and technical criteria and descriptions is necessarily accompanied by a high degree of abstraction and formalization,

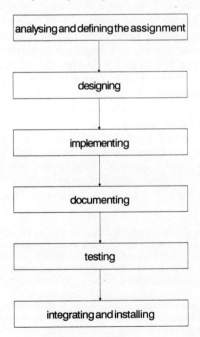

Fig. 4.2. Development stages in software engineering.

moulded, more or less consciously, by designers' interpretations, values, and social orientations. Depending on the development technique employed, similar transformation problems may already occur at the definition stage. This may be so particularly when a largely operationalized specification is itself part of the assignment.[*]

Transcending traditional assignments and disciplines is much more difficult for social scientists and scholars concerned with humans.[†] Social scientists in particular, analytically oriented as they traditionally are, lack the drive to shape. This is not merely the result of a pragmatic division of labour among academic disciplines. Rather – as the example of the critical theory bears out (Adorno and Horkheimer 1969) – it is more or less constitutive of social science. Interdiscplinary and participatory technology shaping thus poses a fundamental challenge to social science. The idea of a shaping-oriented – and hence constructive – social science is, however, increasingly gaining ground (Rauner 1988a). Social science complements engineering science by tackling the technical implications and effects of social shaping criteria and requirements, for instance with regard to computer-aided work systems. Any abstract sociological concept of technology fails to recognize the latter as the object of the shaping of work and work systems. True, such a concept sees technology generally in its historical dimension as mediating between humans and nature, or, sociologically, as, for instance, objectifying labour. Yet the concept fails to consider technology's technico-instrumental qualities and quantities.

Crossing the border requires at least those scholars concerned with humans and those social scientists who see themselves as shaping-oriented to acquire sufficient technical knowledge to be able to engage in constructive shaping-oriented communication with technologists and engineers.

True, crossing the border in the course of an integrated and participatory development process is a serious academic problem, as these few observations bear out and as we have shown in Chapter 3 in particular. Yet crossing the border is far more important to any effective cooperation among users, and between the latter and academics. In order further to clarify the concept, we need to distinguish between "employers" and "users". Only then can we define at what point the practitioners at company level need to transcend their conventional scope of acting.

Employers, according to Koslowski (1988), are the representatives of the heads of organization or organizational unit (such as company owners, management, middle management). Employers are therefore those who "actually commission" a given development. Their supreme object is to enhance their company's productivity. Initially, they are not primarily

---

[*] In the HC-CIM project, development resulted in a comparatively abstract and open description of the HCCIM system to be developed. Therefore the project lends itself particularly to an analysis of transformation processes. The Data Management Group attempted explicity to organise transformation as a shaping process. They did so by turning the underlying values and objectives into a detailed work scenario, thus making them available to everybody involved.

[†] Academic disciplines concerned with humans, such as medicine and educational studies, are far more explicitly oriented towards constructive solutions. Both disciplines are action-oriented and derive their ultimate legitimation from their ability to cure people, or to impart to them knowledge, skills and attitudes.

concerned with humanizing working life nor with the social shaping of technology. At best, they tentatively adopt these shaping orientations under the combined pressure of statutory requirements, union agreements and company staff's right of co-determination. Transcending this restricted shaping interest involves, among other things, taking note of users' shaping interests and making full use of shaping possibilities and development strategies in cooperation with science and research. Such transcending of traditional employers' interests is promoted by the recognition that, even if the primacy of enhancing productivity is maintained, there is more scope for the human-centred and socially compatible shaping of technology than has hitherto been supposed. Quite a number of authors deem the humanization of working life and human-centred technology to be central to more effective production.

Users, by contrast, are those wage-earners who in their capacity as skilled workers, specialists, or designers (in the case of CAD) have, more or less directly, to operate, service, and if necessary repair computer-aided work systems. Users' shaping and interest is traditionally related to factors immediately relevant to their work, such as working conditions, forms of work, job content, stress, wages, etc. Taken together, their attitudes and abilities are geared towards fending off and "defending". Crossing the border involves, here, too, learning to articulate one's work-related needs in the form of interests, translating them into criteria relevant to shaping, and communicating about them with other agents, and that beyond traditional attitudes, abilities and activities. This is, moreover, an important precondition for understanding and assessing the other agents' – including employers' – interests with regard to shaping.

Attitudes, abilities and activities (see Fig. 4.1) are helpful categories for describing and classifying those concepts and strategies that are suggested for the participation of the people affected by technological innovation. Opperman (1983) distinguishes between the degree and the intensity of the conflict attendant upon articulating one's interests. By reducing the many facets of "shaping competence" – which must be seen as a potential that needs bringing out – to the representation of interests, his classification scheme remains confined to the limits prescribed by the classical forms of "participation" and co-determination.

4.1.3 Shaping Competence

Elaborating upon the concept of crossing the border requires clarifying the term shaping competence. Traditionally, the competences attributed to the agents involved in the development and implementation of tchnology have been seen as very narrow and specific. While knowledge specifically relating to technology and its development has been ascribed to designers, assignment-related knowledge (definition of objectives and purposes) has been attributed to employers. Users have usually been afforded opportunities for having their interests represented according to their company's industrial relations code and the right of co-determination.

Attitudes

All agents are equally experts of their daily lives. This means that, in the development of computer-aided work systems, the agents involved have to become aware of their respective expert status according to their place in the social hierarchy. Only then will they be able to understand and assess the other agents' expert status. Learning to understand and assess the limitations of one's own experience, knowledge, abilities and interests is an important precondition for understanding and accepting the other agents' attitudes, abilities and actions. This is not tantamount to giving precedence to others' interests over one's own, but to rendering the latter amenable to free and uninhibited discussion. Lack of a considerable degree of intellectual curiosity in other disciplines' approaches, methods and results precludes any productive communication among scholars from different disciplines.

The fact that designers (engineers) have a professional interest in shaping is evidenced by their profession. This professional interest in shaping has a marked instrumental quality. Extending it into taking the social shaping perspective into account is an important step beyond the limits of engineers' traditional design interests. This extension implies engineers' readiness to take note, and to evaluate the technical implications of the desires, experiences, judgements as well as the ability of those who are affected by a given development process. The individual interests and abilities of the people concerned form the ultimate starting point in any shaping of technology. Such an attitude is alien to, among others, those scholars who, at best, consider the people concerned from the point of view of acceptance or who turn them, in the shape of objectivistic personality models, into objects of their research. In an integrative and participatory development process, social scientists need to divest themselves of an attitude that turns other agents into the objects of their discipline. This requirement poses a considerable challenge to social scientists, who, in accordance with their professional self-image, are regularly tempted to assume the part of moderators and "meta-scientists" in such situations and to define the latter in their favour.

Within the concept of crossing the border, communication is not a sociological method but an open interaction among subjects only, who jointly define their situation and the rules governing their communication. As part of their communication, agents may also agree to take turns in playing the parts of subjects or objects in the joint development process. Users as well as social scientists and scholars concerned with humans can be assumed to possess only a very limited interest in, and motivation for, shaping-oriented cooperation. The high degree of plasticity of IT and its use for objectifying, controlling and organizing mental and social processes has led to a marked change of interests. The project of co-determination is increasingly seen as a reactive, one-dimensional "shaping concept", highly contested between those in power and those opposing them. This project needs extending into a participatory shaping concept. This is not tantamount to a linear extension of co-determination, but to the moulding of the "productivity and social relations pact" (Hildebrandt 1987).

The concept of integrative and participatory shaping of work and technology incorporates both workers' co-determination and the quality circles engendered by management. Both thereby assume a new quality guided by interests.

IT has led to dramatic changes in the object of technology shaping. The possibilities for mechanizing mental and social processes, the development of totally novel information tools, and the new potential for multiplying political and economic control have induced many social scientists to venture the transgression of the conventional boundaries of their discipline and to turn to the issue of shaping.

Abilities

Changes of attitude imply different and new abilities that are indispensable to the crossing of borders towards technology shaping. Here, we can only briefly exemplify in which direction agents have to enlarge their abilities beyond traditional professional and job-related boundaries in order to meet the demands of a holistic shaping concept.

Agents' groups themselves display limitations and restrictions. In the context of company- and concern-specific as well as national traditions, there evolve specific differences not only in attitudes, social and cultural orientations, but also, and markedly, in abilities. Differences in the production concepts found in the trades as compared to industry, between different branches, between companies within the same branch, and between industrial nations clearly express the differences of qualification to be found among staff. Existing qualifications are usually conceived of as linear requirements that are more or less determined by the development and implementation of technology. It has only been recently that close comparative studies on an international scale have shown that the scope of qualifications in the trades and professions is not grouped around a structurally determined core (with the qualification structure being, on principle, technologically determined). Instead, there exist marked structural differences. The pronounced orientation towards computer science in the training of US engineers employed on computer-aided work systems differs markedly from the orientation towards machine-tool building that predominates in the FRG. Even more pronounced are the differences in skilled workers' qualifications. Their proportion in the workforce, their training, and their abilities differ markedly in comparable companies in West Germany, the USA and the UK.

Crossing the border therefore means, first of all, to reject the determinist concept of qualification requirements and to become aware of the relative nature of existing abilities. The shared knowledge that work and technology are shapable, the ability to think in alternatives, and the fathoming of shaping scope are equally important to all the agents involved in a holistic shaping process.

"Crossing the border requires all agents to have a profound knowledge of labour and production processes, including possible solutions in the area of work organization and their technical implications. On this basis alone job-related solutions can be realized which assign technology the function of means" (Ehn 1988).

If, by contrast, professional designers' technical knowledge and know-how dominate the shaping process, solutions will tend to be technocentric. The other agents will be prevented from using their specific abilities in designing technology. This is because such dominance will reduce to constructional scientific methods the mediation between the different jargons and forms of expression employed by the agents involved. Apart from the mutual enlargement of knowledge as an important precondition of a holistic shaping process, social abilities, too, are central to the success of cooperation and communication. In order better to assess and accept the other agents' abilities and interests, one has to conceive of and assess one's own abilities, interests and orientations as conditioned by society and the course of one's life. As a result, one realizes their limited usefulness to the shaping assignment at hand. This insight is a key ability which needs promoting. Moreover, one has to be able to put oneself in the other agents' place in order to think about a problem from their point of view. Also, and relatedly, one has to be able to listen, and not to judge prematurely. These are further important preconditions for the participatory and integrative shaping of computer-aided work systems. Finally, one has to be able to relate one's own abilities, and their limitations, to the extent and quality of the shaping assignment at hand: insight into the limitation of one's problem-solving competence is constitutive of participatory technology shaping.

Acting

The term technology shaping, unlike "designing" or "constructing", is intended to denote a more than merely purpose-related development process which begins with a clearly defined assignment to be accomplished according to equally clearly defined rules. The fact that engineers have been assumed to be guided in designing by a high degree of purpose-relatedness has largely helped to stabilize the use of technocentric methods, which have been common up until now. Purpose-relatedness excludes those dimensions of acting that aim at agents' creativity, at the real world's virtuality and adaptation to a given situation, and at the tacit skills, which are particularly important to the shaping of technology.

Crossing the border requires all the people involved to adopt an experimental basic attitude and to be able to behave experimentally.

The participatory and experimental search for alternative technical solutions extends far beyond the restricted, rigid, scientific experimenting that has been central to empirical cognition since Galileo and Bacon. The dominant concept of experimenting has been moulded by the sciences' conception of nature. Scientists are concerned with snatching from nature its "secrets" at all costs, as Bacon put it, in order ultimately to govern it. Shaping, by contrast, employs an experimental approach whose objectives are largely open and where the people concerned enter into the given situation. Shaping by experimenting is characterized by a communicative structure, curiosity about trying things out, and a creativity that is adapted to a given situation. The exploratory prototyping to be found in software engineering shares many of the features common to shaping by experimenting.

Any participatory shaping process requires agents to make use of new aids in order to overcome the barriers to action that have been put up by jargons and by the difference between academic and everyday registers. Illustrating and visualizing shaping alternatives are particularly relevant to the pulling down of any barriers impeding understanding. Experimental and exploratory simulating and prototyping are adequate means of enhancing agents' communication about the interests, objectives and means relevant to shaping.

4.2 Participatory Systems Development

4.2.1 Hierarchic–Sequential Design Tradition

Both in construction science and at the initial stages of software engineering, there evolved methods of technology development which are guided by purpose-related and planful action. A rigid distinction is made between the translation of development objectives, intentions and purposes into a specification (requirement profile) and the design process proper. In the course of the latter, the requirements laid down in the profile are gradually turned into a systems solution. The definition of ends is conceived of as a normative process occurring prior to design. Consequently, the latter appears to be value-free. VDI (Association of German Engineers) guideline 2222, for instance, distinguishes a sequence of four stages in the accomplishment of a given design assignment:

1. Planning: functional analysis, definition of overall and part functions
2. Conceiving: approximation of conditions of use, solution of material and dimension tasks
3. Designing: designing true to scale
4. Detailing: detail drawings, appliance design, check-list, operations planning and scheduling (see Fig. 4.3).

Initially, software engineering evolved in the tradition of classical construction science. Hence it conceived of software development, by analogy with construction, as the hierarchic–sequential transformation of a specification into a product. This transformation is broken down into the following part abilities, which are arranged in a linear order in terms of a stage model (see Fig. 4.2):

1. *Definition of assignment.* This stage involves analysing problems and assignments unless there is a given specification. Requirement analysis leads to requirement definition in the shape of an operationalized specification of tasks. Quality standards are defined according to which attainment of goals is checked.*

2. *Designing.* In the design process, the specification of requirements, in terms of a description of the desired properties, is transformed into a description which includes the definition of the systems structure, of the data and control flows, and of the distribution of components. This transformation is the key event in the design process because, here, the purposes defined

---

*Balzert points to the limits inherent in this first development stage: "It is impossible to define in advance the total behaviour of a complex system in detail. Particularly with regard to systems with intensive human machine communication, it is nearly impossible to predict what people will be doing with the system once they have obtained it" (Balzert 1986, p.110).

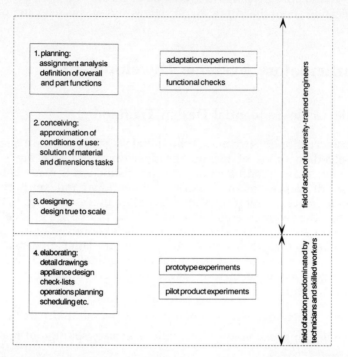

Fig. 4.3. a Hierarchical stages of technology development (according to VDI Guidelines 2222).

are irreversibly translated into technical structure. In classical design, personal orientations, values, interpretations and objectives, all of which have a considerable impact on this transformation, remain largely implicit.

3. *Implementation*. In software engineering, implementation involves translating the design into an operational software system.

4. *Documentation*. Adequate use of the "product" hinges on documentation. This should include:
Specification of requirements
Description of program (description of design)
Administrative records
User manual

5. *Testing*. Testing is traditionally understood as a comparison of nominal and actual functions in order to prove a program to be free of errors. Testing in a broader sense means evaluation and includes personal assessment of the result by the people involved (see Fig. 4.4).

Finally, the five main stages of software development are followed by:
Integration and installation
Maintenance, service, administration

These more or less linear–sequential methods are often coupled with a

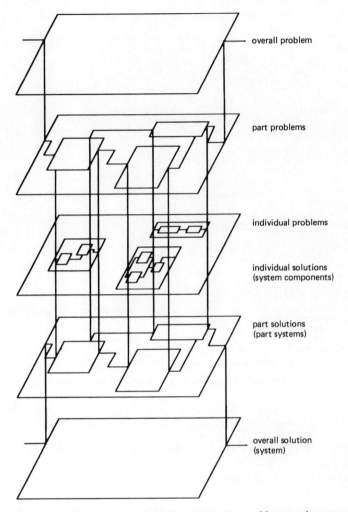

Fig. 4.3. b Method of analysis and synthesis for structuring problems and systems.

hierarchical top-down structure in decision making and in the labour process (see Fig. 4.5).

This widespread structure of design is characterized by a hierarchical division of labour and control, which has proved less productive in those development projects that are intially not amenable to definition in terms of their use-value.

These classical methods of technology development, which are applied to the hierarchic–sequential concretization of a given specification (see Fig. 4.2), dominate software development, too. The division of development into clearly defined stages has been intended to render the process transparent and improve project control. In stage models, this is achieved by clearly defined intermediate results that are evaluated and checked at intervals by project management. As a result of these checks, the subsequent development stage is tackled, or else the intermediate product is revised and amended.

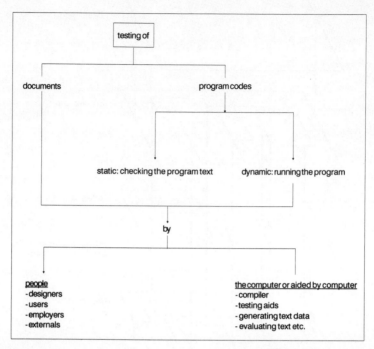

Fig. 4.4. Testing of software documentation and codes (according to Koslowski 1988).

Koslowski points out the reasons why stage models have become so widespread:

1. By being divided into stages, each resulting in intermediate products, the development is structured and easy to grasp.
2. This considerably facilitates planning, organizing and checking dates and staff development.
3. Systems development is rendered "linear" with checks and, if necessary, revisions of intermediate results making feedback superfluous.
4. Division into stages and corresponding part activities helps emphasize analysis and design (see Fig. 4.3). The "product" aimed at thus gains priority over the development process.

Koslowski's assessment is valid only with regard to the general appreciation of the formal rationality of these methods and the firm ground they have taken in teaching engineering science. His assessment does not apply, however, to the reality of developing and designing. Here, those methods geared towards purpose-related action are rather secondary. It was not until computer-aided planning and control of development was introduced, to which stage models particularly lend themselves, that the latter gained more recognition in practice.

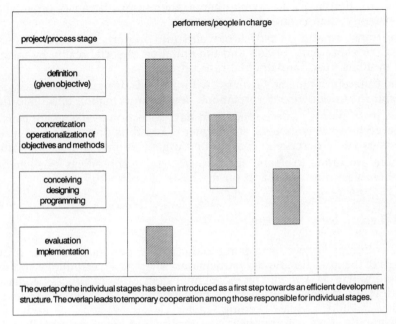

Fig. 4.5. Hierarchical division of labour and control in stage-oriented development concepts.

4.2.2 The Limitations of Traditional Design

As early as 1981, Floyd pointed out that stage models were successful only in special cases that met the following criteria:

1. The requirements of an organization and of the people involved are clear and can be met permanently by a tailor-made IT system.
2. It is possible to state these requirements completely and precisely at the beginning of systems development.
3. Recording these requirements by means of written, semiformal documents poses no problems.
4. The different levels of description and concreteness have been transformed correctly, i.e. without losing or distorting any information (Floyd 1984, p.135).

These assumptions obviously apply but to a limited class of assignments of little complexity.

The reasons why stage models frequently fail, are increasingly rarely applied or are not very helpful in finding adequate solutions have been summarized by Koslowski (1988) thus:

1. Complex organizational structures, particularly when based on possibly heterogeneous interests, cannot be described correctly, completely and unambiguously.

2. It is impossible to describe organizations over an extended period of time and to derive static requirements therefrom, because organizations constantly change and develop further.
3. The transformation of aspects of reality into properties of technical systems and corresponding models and the imitation of social reality by means of IT systems pose fundamental difficulties.
4. The different linguistic "realms" within which designers and users move result in serious misinterpretations, loss of information and distortions.
5. The instruments of description and data collection available are primarily geared towards engineers' requirements and thus do not lend themselves to adequately describing complicated organizational and social relations.
6. There are limits to users' ability to recognize the effects of systems on themselves in advance and on the basis of abstract plans.

4.2.3 Technology Shaping

These criticisms largely correspond to our own experience and underpin our concept of the interdisciplinary participatory shaping of technology and work. The alternative to a hierarchic–sequential development process consists of regarding the stages as functions of development and in fulfilling them in a parallel fashion, while keeping them closely linked to each other. Consequently, they continuously presuppose and bear on each other. Such rolling development (see Fig. 4.6) takes into account four important aspects:

1. Objectives may be changed as a result of integrated development. Such changes are simultaneously learning processes for all the people involved.
2. Employers and users participate in their capacity as experts of their work situation.
3. The development product is seen as one version of the system.
   This implies giving priority to technology as a process over technology as a product: technology as an adaptive element of organization development. Adaptability thus turns into a central technical systems property or an important shaping criterion.

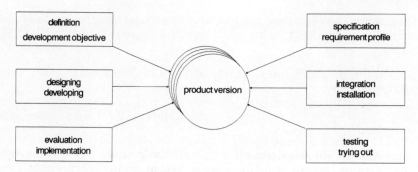

Fig. 4.6. Rolling system development.

4. Development is no longer a linear process. Rather it is seen as a sequence of objectives which result from the interplay of producing, testing and revising/adapting. Thinking in terms of versions is particularly adapted to (a) the plasticity of IT and (b) the permanency of organization development.

Rolling systems development within the human-centred CIM project for developing individual parts follows the pattern of a "generative conduct system" (Aregger and Frey 1975).

The concept of "rolling technology shaping" corresponds to techniques of process-oriented and evolutionary systems development that are gaining increasing importance in software engineering. Both concepts (a) feed back and branch out between the various activities involved in systems development and (b) work with preliminary systems or systems components that can be realized easily and quickly and be modified. Ideally, there exist operational preliminary systems variants at each moment of software development and implementation. Process-oriented methods of systems development which involve employers and users in the development process through fast preliminary realization of systems and systems components are classified as prototyping:

Exploratory prototyping

Experimental prototyping

Evolutionary prototyping

are seen as alternatives to traditional design techniques. They are adequate to the systemic and IT character of "new" technology, are more efficient and may make a considerable contribution to rendering technology shaping more democratic (Koslowski 1988, p.168).

Exploratory prototyping aims to support communication between designers and users. Alternative solutions, which afford direct experience of important aspects of the system to be realized, enable (potential) users more adequately to assess the changes in their work consequent upon the introduction of computer-aided work systems. The problem solutions enable users to develop tangible ideas. Exploratory prototyping is of strategic importance at the transition from requirement analysis to design. Very early on, the immediate experience of operating the prototype version enables customers and users to render their objectives more precisely and, if necessary, to revise them in a way that is always close to real conditions.

Experimental prototyping presupposes a comparatively stable objective or requirement description and helps adequately to realize the existing specification. Here, too, prototyping encourages participatory development: the "experiment" considerably facilitates designers' and users' grappling with, and assessment of, potential solutions.

Evolutionary prototyping is the furthest reaching kind of prototyping. It assumes that technology development can only adequately be conceived of as integral to the development of constantly changing organizations. It requires IT systems that can permanently be developed further. "This implies the total rejection of stage concepts, because the boundary between systems development on the one hand and service and maintenance on the other has vanished" (Koslowski 1988, p.171).

Possibilities and Limits Inherent in Prototyping

It is for good reasons that prototyping has gained such relevance in software engineering. This trend has been furthered by the high degree of plasticity of software technology, the enlarged scope, afforded by new-generation computers, for the development of prototyping tools and by user participation in shaping "user surfaces", i.e. human–machine interfaces (and that not least for reasons of economic rationality). In classical machine building, by contrast, protyping of the kind described above is ruled out by the physical and objective starting conditions which are taken as given.

On a larger canvas, there is, however, a marked tendency for using computer technology in variable conjunctions with other media for representing potential realizations. The architect group Wickfors in Uppsala (Sweden), for example, has developed computer-aided techniques for representing building designs from various perspectives:

Integrating the particular building into a whole range

The wider surroundings

The angle of the main entrance

Bird's-eye view

Various interior views

This type of representation does not render obsolete the realistic models that have always been used in urban planning and architecture. Furthermore, this technique allows dynamic representations, for example from the point of view of a pedestrian or a car going by. The participation of citizens in urban planning or, generally, communication between architects and their clients, or the people concerned, thus assumes a new quality.

Like prototyping, dynamic representations of more complex processes, such as factory logistics, have become similarly important. The intensity of participatory technology shaping hinges upon the quality achieved by the representation of the objects under development (in the widest sense) for the people concerned. These representations include social ones, such as decision, planning and role plays.

With regard to CIM, these forms of participatory technology shaping are embedded in a wider concept of socially compatible and realistic technology shaping. The following section documents design instances.

4.3 The Shaping Process

Below, we shall design a scenario of the participatory and interdisciplinary differential shaping of computer-aided work systems. This scenario will be based on both our general reflections upon the social shaping of work and technology and our actual project experience. This shaping concept has to be seen as one dimension of organization development. Differential shaping means that, in the process, there have to be different shapes to choose from. Choice may occur in:

The transformation of technological basic innovations into applied technology (e.g. computer-aided work systems)

The organization of work (Ulich 1978)

The participation and cooperation between academics, users, and the people affected.

Particularly those immediately affected need the opportunity to choose forms of participation that accord with their specific capability to express themselves and their abilities and interests. In several important ways, this concept differs from the attempts at designing anthropocentric technology hitherto tried out and presented. Three approaches to realizing the concept of human-centred technology shaping can be distinguished:

1. In the engineering tradition, the search is for objectively valid shaping criteria or reference models. These are to serve as objective criteria of design. Once found, they are expected to help distinguish between humane and inhumane technology without recourse to the arduous processes of the participation of those affected and the interdisciplinary dialogue. Due to the narrow limits within which the concept of human-centredness can be objectively operationalized, this approach is hardly relevant to the development of computer-aided work systems (Rauner et al. 1988a).

2. Alternatively, one can go on optimizing procedures that have proved to be useful in the interdisciplinary and participatory shaping of technology until defined procedures are finally arrived at. With regard to software engineering, a large number of such procedures have been suggested. This approach offers very limited possibilities for bringing individual abilities, interests and experience to bear on the shaping process as required by a given situation and beyond defined procedures. These limits are, however, far less confined than is the case with criteria catalogues and reference models. The search for an approach to the shaping of anthropocentric technology that does not turn those involved (including those immediately affected) into subjects of the shaping process leads into a dilemma. All the people involved can theoretically be the subjects of the shaping process, provided that there is a maximum scope for acting, that the situation is objectively completely open and does not imply any assumptions. Those involved may, however, experience such an open situation as highly restrictive, with the lack of any orientations to which action can be geared potentially leading to inability to act. Keeping the balance between the two shaping principles that contradict, yet presuppose, each other – a maximum of planned and purposeful proceeding on the one hand, and large scope for acting and open acting situations on the other – is a challenge that has to be met anew again and again.

3. The third approach, which offers a practical way out of this (theoretically inescapable) dilemma, consists of rule-governed cooperation of academics and practitioners. While, in the first case, shaping occurs at the level of products, and, in the second, at the level of defined processes (procedures), the third approach is based on rules of dialogue and cooperation between those involved in the shaping process. If the object of shaping is humane work in computer-aided and -integrated manufacture, then this approach alone can lay the scientific base for the shaping concept (see "Shaping Paper" and Chapter 1).

This final chapter is therefore intended:

1. To present and give reasons for the rules of a participatory and interdisciplinary approach to shaping.
2. To exemplify by means of scenarios how these rules may be translated into actual processes of integrated technology shaping.

The concept proper of the anthropocentric shaping of technology is thus confined to these rules. The examples given of the possible application of these rules are intended to stimulate thought and to demonstrate how they can be transformed into interdisciplinary and participatory shaping processes. The examples chosen are not mere figments of our imagination, but are doubly related to the practice of our project. They represent typical cases that have been turned into scenarios and transcend the practice tried out in the project by:

1. Being summarized into linked-up shaping processes.
2. Indicating possible branchings off, which are typical of a differential shaping concept.
3. Taking account of negative project experience in that, in the cases presented, this is given a positive turn.

Rule-Governed Shaping of Computer-Aided Work Systems

1. Any shaping process is based on an assignment that is clearly formulated in a manner comprehensible to all (employers, academics and users) and on which agreement has been reached.

This rule does not define the level of abstractness at which the common assignment is formulated. It is, however, important for the assignment to be so clearly phrased as to be understood by all and to serve as both an orientation in the problem-solving process and a criterion for the evaluation of (intermediate) results. According to this concept, however, the assignment is not rigid in the sense of a specification, but allows a rolling shifting of objectives. With any participatory and interdisciplinary shaping process also representing a special kind of learning process for all involved, one has repeatedly to make sure whether or not the assignment and the concomitant working practice anticipated have to be modified, revised and/or specified on the basis of enhanced insights.

In the project on human-centred CIM systems, for instance, the assignment was formulated at a high level of abstraction due to the pronounced (desired) heterogeneity of project participants. This and the general lack of experience with interdisciplinary and participatory research and development projects formed the major difficulties in gradually transforming the basic idea of human-centredness into technical artefacts. The concept of gradual or rolling assignment development and shifting of objectives is thus a response to the methodological deficits shown by the project on human-centred CIM systems. Rule one implies the demand for formulating the assignment as concretely as possible. If no agreement about the assignment can be reached at an advanced level of concreteness, due to lack of knowledge or difference of interests, if, however, general interest in fulfilling the assignment in general persists, then it is usually advisable to talk about the assignment at a more abstract level. This, however, must not happen at the expense of clarity and comprehensibility.

The transition between levels of assignment description is, in the case of concretization, linked to a process of discussing further-reaching aspects of the objective conditions to be concretized and assumptions to be made. This, however, already forms an essential element in the shaping process itself. Formulating the assignment thus involves choosing and assessing the adequate level of concreteness as well as defining the distance between the concrete project result and the chosen level of assignment description.

2. At regular intervals, particularly at the end of defined work stages, the project group reflects upon the method employed and the quality of the intermediate result achieved, deciding form and content of the subsequent project stage.

The second rule is characterized by the fact that there exists no part authority in the project, such as might be represented by individuals (project leaders) or by members of certain disciplines who take responsibility for the method employed by virtue of their profession. The project group as a whole determines the content and form of subsequent processes. In doing so, it uses the existing rough schedule and the suggestions to be made, if need be, by the steering group (see rule 4). The alternation between action and reflection in action or, in other words, between communication and meta-communication is an important rule in any shaping process which all project participants together are the subject of. This rule precludes the search for linear methods of design. Ongoing reflection throughout the project in conjunction with branching and open shaping processes are central to the overall concept of rule-governed development. Stages of reflection and meta-communication have to be integrated into the overall process so as to operate to the project's full support. Otherwise, shaping projects stand in danger of evaporating into meta-communication and of failing eventually. Practising and applying rule two should be subject to time-limits agreed upon in advance by the project group.

3. The groups involved have the opportunity and are encouraged to contribute to the shaping process those forms of expression that correspond to their experience, their abilities, interests and wishes.

The concept of working jointly on assignments beyond the boundaries of disciplines and fields of practice requires understanding and communication as well as media for transmitting knowledge, designs and experience. Here, again, tensions exist. Fulfilling a technical assignment commonly presupposes technico-scientific instruments. This is usually part of engineers' and engineering scientists' professional competence. If this recognition were to give rise to the rule that the shaping of technology has to occur in the jargon and the medium of the technology and technical science at issue, this would erect insurmountable barriers to the people involved and affected who, for good reasons, are equally seen as experts of the working practice to be shaped. Each of the expert jargons and methods is simultaneously a barrier to, and a necessity of, communication and participatory/interdisciplinary shaping. This applies equally to academics, and the jargons specific to their disciplines, and to skilled workers and managers. Thus the shaping process has to allow for stages at which group-specific media and the means of expression common to the people involved are used as well as for stages at which group-specific ideas are translated so as to become amenable to being discussed and incorporated into the overall result. Scenarios have proved to be a particularly effective form of expression at the integrated stages of the shaping process. With every project on the shaping of computer-aided work systems requiring changes in, for instance, one sector of the factory, the description of the events anticipated at company level, such as the processing of an order, is an excellent form of communication. Designing such scenario "pictures" about the change desired has proved successful in our project. This allows all the people involved (a) from their perspective of experts to participate in designing future work and company events as a whole, and (b) to use the jointly designed scenario as the complex base for translating the above into the perspectives of the various groups affected.

4. Representatives of the people involved in the project with their specific abilities, experience and interests make up a coordinating group which has two main functions:
(a) Integrating and systematizing the contribution to the project made by the groups involved
(b) Organizing "double loop learning" or meta-communication (Habermas 1981) in order to liven up the process of self-evaluation.

If the size of the project group exceeds a certain number and if the project is also internationally composed, as with ESPRIT, then the setting up of such a coordinating group is necessary for practical reasons alone. Here, the aspect of representation is not a step towards establishing a hierarchical project organization, but the result of the principle that both moderation of the project dialogue and cooperation have to occur collectively. Our (the Social Science Group's) own practice of drawing up and discussing the "conceptual framework for research and development projects in the area of CIM" (see Chapter 2) has shown clearly that, by assuming the role of moderators in the discussion of this concept, we as social scientists were unable to overcome the barriers to dialogue erected by the framework that was itself drawn up in the terminology of our discipline. Social scientists ought not to act as the moderators in the interdisciplinary and participatory project practice as a matter of course. Their seeming professional suitability for this task often turns out to be a barrier to communication. Assigning the parts of moderators to social scientists usually impairs the problem-solving process. Moreover, the social situation is informally defined by the jargon of social science. Thus in these cases, contributions by social scientists often dominate the project situation without in the least advancing the process of technology realization. The rule we advocate conforms to a position in activity research which maintains that the opposition of objects and subjects in the research process has to be abolished. Setting up a coordinating group for the moderation of shaping projects is an important step towards realizing this basic rule of research methodology. The coordinating group plays an important part in initiating the formulation and progression of the assignment and the explication of problem-solving horizons of the groups involved (as a reflection upon intellectual barriers).

Referring to our own project practice, we shall translate the concept of the rule-governed participatory and interdisciplinary shaping of technology, as established above, into two alternative shaping processes. These processes, though following two different basic patterns, are not presented and discussed with a view to establishing which of them is superior. Any decision on either of these approaches is completely up to the people actually involved in the shaping process. Of course, both methods may also be combined with each other.

This basic model of participatory and interdisciplinary differential technology shaping (PIDTS) is characterized by an alternation between the states of specialization by discipline- and interest-specific groups and generalization by groups of mixed composition. This is an iterative method, because the two stages may be passed through several times in succession depending on the distance between the assignment and the concrete project result (see Fig 4.7). Rule three in particular suggests the mutual translation of specific interests and of experiences formulated from the specific point of view of those affected into generalizable and integrated aspects of shaping.

The four-stage method (see Fig. 4.8), on the other hand, follows a different logic while also being based on the four rules of PIDTS. In content, this four-stage method is based upon the six dimensions of technology shaping (see "Shaping Paper") as established in Chapter 2. Methodologically, it begins in two stages from a concrete work practice that has been found to be in need of change. This is then specifically represented by the groups of participants and assessed with regard to the six shaping dimensions. Subsequently, these different representations are integrated into one picture of the practice in need of change. The stages of the representation, analysis and assessment of existing practices are followed by two further stages. These involve transforming the results of the analysis into the scenario desired and its subsequent concretization and translation into the perspective of each of the participant groups involved and affected. All of these four stages together can be seen as one sequence in the designing of a factory scenario, for instance, and thus as one stage in the overall process of PIDTS.

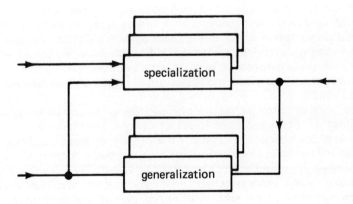

Fig. 4.7. Iterative two-stage design method.

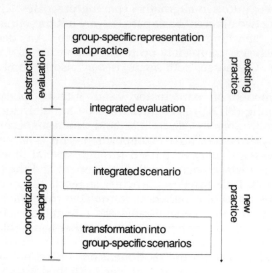

Fig. 4.8. Four-stage method for designing a scenario.

4.4 Prospective Example of the Shaping Process as a Rolling Development

The principle of rolling development may be practised in a variety of ways. Based on both positive and negative experiences within ESPRIT project 1217 (1199), this section offers a description of one of many possible methods of rolling development. It should be stressed that the following example is not the model for the shaping of CIM technology from a human-centred perspective. Rather, it is a prospective case for illustrating how the principles of rolling development may be carried out in an interdisciplinary context.

The following example presupposes a rather complex task; i.e. it is more than just a prototype of a tool or software system for a stand-alone human–machine system. In the following case the question which is addressed concerns how the organization, technique and educational aspects of a factory (or the production centre of a manufacturing company) may be viewed as an integrated entity.

Neither organization, technique nor education has an ontological status. That is, they all open from the onset of design. For example, we consider CIM as one of the possibilities and not the definite objective of design. In this sense, the following example of rolling development is more wide ranging and open than the objectives of the ESPRIT project. The essential point is to establish a shaping process which, from the outset, includes a dialectical interaction between practically experienced people and engineers, towards the shaping of human-centred manufacturing.

4.4.1 Preparation of the Shaping Process

We suppose that at least three project design groups should be established: a group of end-users (U), a group of social scientists (S) and a group of engineers (E).

Based on the experiences of the Danish group within the ESPRIT project, the group of end-users may be chosen from within the same company, although not necessarily from the same department. Also, it should be people from the same level in the company hierarchy to avoid introducing intra-company politics into the group's design work.

The social science group may include people from different social science disciplines (e.g. psychology, sociology and ergonomics) but members should be encouraged to dispel any disciplinary parochialism or narrow-mindedness.

The engineering group may include people from different engineering disciplines in order to cover the different problem areas within the project. It may also be an advantage to include engineers who have some experience in CIM system implementation and education in a practical setting.

Finally, there should be sufficient time to discuss the meaning and possible contradictions surrounding the concept of human-centredness both within, and between, the groups, in an effort to encourage the formation of a mutually agreed definition of, and philosophical basis for, human-centredness.

Each of the groups should choose one of its members as their representative on a coordination group (CG). This group of three has a number of important functions to perform during the design process (see rule four, section 4.3). Before the groups begin working, the CG defines the tasks and the problem-solving horizon of the project in order to secure comparative standards across the groups. These standards must be agreed by the three groups.

In order to secure both rolling development and to prevent a total lack of coordination and focus, the process should be governed by the rules outlined in section 4.3

4.4.2 Future-Creating Workshops

During the initial stages of the project, it may be fruitful to carry out a number of rather open and loosely structured discussions. "Future-creating workshops" incorporate both a critique of existing conditions of work and technology and creative ideas concerning new conditions and suggestions for the realization of human-centred work and technology.

In the ESPRIT project, the Danish CAD group gained experience in using this method. The social scientists held three future-creating workshops with three separate user groups. Although a number of interesting ideas were produced in these workshops, we recommend the following procedure for further workshops.

Each of the three groups (U, S and E) separately describe their initial versions of an organization scenario or equipment prototype in their preferred language (e.g. written word, physical models, drawings, role playing). The process is illustrated in Fig. 4.9. There are several reasons for separating the groups according to common language and experience at this stage of a project. First, although the Danish user groups produced a number of interesting ideas

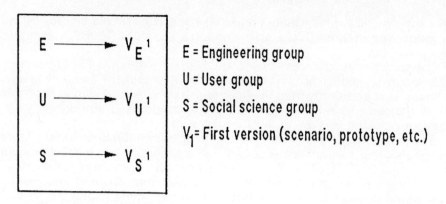

Fig. 4.9. A future-creating workshop.

during the ESPRIT project, their work was directed and influenced by questions and stimulation from the social scientists. The problem setting was devised by the social scientists and the engineers were not involved at all at this stage. A much wider set of design options and ideas may have been produced if the three groups had been working in parallel at the same task using their own preferred methods of discourse and enquiry (rule three in section 4.3).

4.4.3 Shaping Workshops

Depending on decisions made after the future-creating workshops (see rule two in section 4.3), the process may continue in different directions. In this case we suppose the next step to be a "shaping workshop" of mixed groups; that is, groups each consisting of at least one social scientist, one engineer and one end-user. Compared with the future-creating workshop, the shaping workshop is firmly structured from the beginning as the following procedural example indicates:

Step 1. The aims and rules of the workshop are presented by a person outside the group who has been chosen to guide the workshop.

Step 2. A user representative presents his or her results from the previous future-creating workshop (V_{U^1} in Fig. 4.9) to the shaping workshop participants.

Step 3. Discussion of the technical and social implications of V_{U^1}.

Step 4. The social science representative presents the results from the social science future-creating workshop (V_{S^1} in Fig 4.9) to the group.

Step 5. Discussion of the technical and practical implications of V_{S^1}.

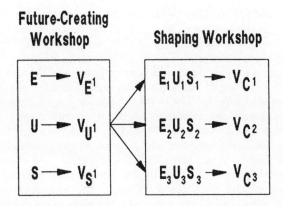

Fig. 4.10. A shaping workshop.

Step 6. The engineering representative presents the results of the engineers' future-creating workshop (V_{E^1} in Fig 4.9).

Step 7. Discussions of the social and practical implications of V_{E^1}.

Step 8. Comparison and discussion of the similarities and potential contradictions, or differences, between V_{U^1}, V_{S^1} and V_{E^1}.

Step 9. Integration (as far as possible) into a common version (V_{C^1}).

The process thus far is illustrated diagrammatically in Fig. 4.10.

Hence we see that the mixed design groups (E_1, U_1 and S_1; E_2, U_2 and S_2; E_3, U_3 and S_3 in Fig. 4.10) are multidisciplinary in order to enable the translation and transformation of the design options derived from the separate future-creating workshops into a common design option or specification which includes technical, social and practical considerations. In this case, the products are three more or less different versions (V_{C^1}, V_{C^2} and V_{C^3}) produced by three parallel multidisciplinary working groups.

At the time of writing an earlier draft of this book, the authors had no experience in running such a workshop. It was therefore decided that, in order to reflect theory in practice, a shaping workshop should be organized around the concept of systems architecture. This absence of an agreed definition and articulation of this concept, it will be recalled, had created a number of problems during the middle period of the project. Hence, the workshop was to serve a dual purpose. On the one hand, we wished to try out a specific kind of interdisciplinary communication, and on the other hand we wished to make further steps towards developing a reference model for a human-centred CIM architecture.

The shaping workshops, held in Berlin during the last year of the project, comprised three social scientists and four engineers from the ESPRIT project, plus two external engineering scientists. The latter were invited in order for new ideas to be injected into the project. Moreover, the presence of two outside experts was expected to help remove the experimental nature of the workshop from the project routine. This seemed important for transferring and generalizing the workshop findings and conclusions from the relatively narrow context of the project.

The shaping workshop was organized around dialogues similar to those detailed above. In the course of the first dialogue, the technical implications and preconditions presented by the factory scenario in the "Shaping Paper" (section 2.4.2) were analysed. In a subsequent dialogue, the engineers presented technical descriptions of a human-centred CIM architecture. These were analysed by the social scientists with regard to their social implications, presuppositions and effects. Three intermediary stages between the factory scenario and systems architecture were also considered (namely work organization, qualification and CIM building blocks) to aid the development of a five-dimensional reference model as well as the gradual transformation of social into technical descriptions and vice versa.

The Berlin workshop was not fully successful in achieving its dual aims, primarily owing to the difficulties faced by the engineers in describing a technical specification of human-centred CIM architecture. This specification was to be developed more fully by the Data Management Group within the project who, in the course of a much longer working process, achieved a level of interdisciplinary dialogue comparable to the communication occurring within the limited time frame of the shaping workshop. More specifically, the data management reference model was complemented by a correspondingly detailed work and factory scenario. The latter served to mediate between the general factory scenario of the "Shaping Paper" and the representation of a detailed technical specification for a human-centred CIM architecture.

The concept of interdisciplinary shaping was thus doubly justified and the consideration of both the Berlin workshop and the workings of the Data Management Group allowed us to suggest how shaping workshops may be developed and organized with far more confidence than would otherwise have been the case.

4.4.4 Coordination of the Results of the Shaping Workshops

Returning to our ideal-type rolling development process, the coordination group (CG) should receive the three "products" of the shaping workshops (i.e. V_{C^1}, V_{C^2} and V_{C^3}). This is followed by discussions and decisions are then taken regarding how to continue the process. In this case, we suppose that the CG produces a design option based on the juxtaposition and integration of these three products.

In the ESPRIT project, a group consisting of three social scientists produced a synthesis of concepts and principles and a factory scenario (see the "Shaping Paper" in section 2.4.2). Although proving to be an important step in the development of the project, this paper was limited in its effect on the overall design process (as pointed out in Chapters 2 and 3).

In order to learn from the project, we suggest that not one but three versions are created. Furthermore, the multidisciplinary coordination group (and not social scientists) should take part in the formulation of a first common design version or specification which synthesizes, or juxtaposes, the possible contradictions between the three earlier versions formulated within the shaping workshops (this process is illustrated diagrammatically in Fig. 4.11). Moreover, the CG may organize an evaluation procedure in an effort to improve methodology and overcome conflicts and misunderstandings.

4.4.5 The Second Shaping Workshop

In this case, we suggest a rearrangement of the groups back to their original format (i.e. an engineering group, social science group and end-user group). The reason for this change relates to the specific task of the second shaping workshop; namely, the comparisons of the coordination groups' deliberations and common version with the original versions produced by the groups in the future creating workshops (V_{E^1}, V_{U^1} and V_{S^1}) in order to criticize and produce new ideas. These are delivered back to the coordination group who develop a new synthesis or juxtaposition (S_2 and J_2). This process is illustrated in Fig. 4.12.

4.4.6 Seminar with Outside Experts

The refined design specification from the coordination group may be further discussed by the three design groups. Alternatively, a few, carefully selected, specialists may be invited to comment on this specification. In the Danish ESPRIT project group, positive experiences of such a procedure were gained by inviting researchers from Norway and Sweden to comment on and discuss the initial concept of human-centredness. The advantage of outside consultation stems from its distance and detachment from the design process.

In this case, we propose that outside experts are invited to a short series of seminars of this kind to enable the integration of a new specification by the coordination group.

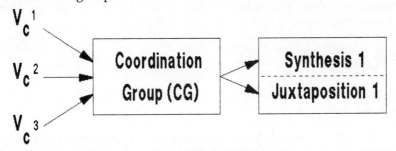

Fig. 4.11. Coordinating the results of the shaping workshops.

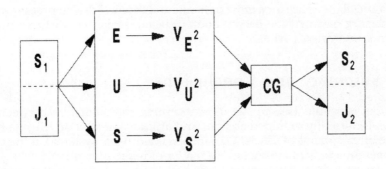

Fig. 4.12. A second shaping workshop.

Subsequently, the procedure is reflected and reappraised according to rule two (see section 4.3). In this case we propose that a third shaping workshop is held with mixed groups of engineers, social scientists and end-users. The objective of this workshop is to prepare an exhibition or workshop for the public. This public airing of the design specification (presented in the form of models, film, prototypes, etc.) is suggested for a number of reasons. First, it might stimulate the project participants. Ideas communicated thus far in the design process are at a relatively abstract level. Through preparation for an exhibition, these ideas necessarily become more concrete. The ESPRIT Danish group prepared a slide show of their CAD work to great effect in this way.

In the present case, it is intended that the exhibition is viewed as a part of the rolling development process rather than as an end product as in the case of the Danish CAD slide show. A second reason for incorporating a public exhibition into the rolling development process stems from the desire to encourage a more open and democratic procedure of technology development. People from outside get a chance to "get inside" the project and even to contribute to it. Third, the project group may get stimulating feedback from such an exhibition if it is shaped in such a way as to facilitate meaningful participation.

Following this exhibition, the coordination group may take the initiative for another internal, self-refletive seminar. Depending on these discussions, the rolling development may again take different routes. In this case, we suppose that some of the ideas are implemented within a company as experimental prototypes. Simultaneously, it may be decided to carry out educational experiments in order to achieve a balance between prospective techniques, organization and educational initiatives.

4.4.7 Conclusion

The above example does not pretend to be *the* model for shaping work and technology. It is an example based on our experiences (both positive and negative) within the ESPRIT project and is therefore rooted in self-reflected practice. However, it should be stressed that, at every step, numerous different directions can be chosen. This example is offered in the hope that it may stimulate designers to break with the dogmatism of a linear, hierarchical process and allow space for a more dynamic and "spiral" design process which has a balance between the participation of users, social scientists, engineers and other interested parties.

4.5 Summary and Conclusions

Our initial questions (see pp. 24–25) concerning the mediation between the industrial–culturally moulded abilities and attitudes of staff on the one hand and the development of efficient computer-aided work systems on the other is not merely – as we have stated – a matter of reflecting and respecting, but also of transcending. The existing forms of work and industrial culture are largely characterized by Tayloristic structures. Set roles and an internalized working practice counteract innovations in all departments and at all levels of the factory, even though such innovation may be in staff's objective interest. The transformation of both management and staff's existing converging interests in reducing Tayloristic structures of work and technology into their personal interests and corresponding willingness to innovate is a key problem faced by those R&D projects that are concerned, directly or indirectly, with organization development. R&D projects with a focus on technology, such as have been sponsored by ESPRIT, run the risk of intensifying old work and factory structures by means of new production technology. Here, collusion may occur between conservative management's interest in maximum supervision and deterministic production control and corresponding staff interest in preserving the status quo.

The experiences and results gained in our project from holistic approaches to shaping point the way towards mediating as well as transcending the traditional orientations and abilities of the people involved. The basic idea of human-centred CIM, precisely because it is not amenable to operationalization, has challenged everybody involved time and again to problematize our technology developments in the areas of CAD, CAM and CAP and their correspondence to the basic idea. Evaluation means, here, continuous communication among all the people involved in their search for a better understanding of human-centredness as truth in the Habermasean sense (Habermas 1981).

This basic idea also illuminates the relation of project planning to routine. The disadvantages of stage models in software engineering also show in the procedures of our and comparable R&D projects. It is from project events occurring in specific situations, from new and unplanned initiatives growing out of learning situations, as well as from the increasing tendency of all participants to see themselves as learners that really innovative insights, experiences and products have evolved. These had not been anticipated and would have been impossible to anticipate.

In the search for new forms of humane, yet efficient, work and technology for the "new" factory, there is a great need for intelligent basic ideas and for committed experimenting behaviour. Conversely, there is less need for computer-aided planning instruments, which reduce agents to objects of their own planning and thus nip innovation in the bud.

A comparison of the national part projects has repeatedly shown the immense importance to the people concerned being the subjects of the development process in their capacity as experts of their daily work and life. The technical results of the part projects CAM (UK) and CAP (FRG) do not transcend the level of solutions already achieved and remain confined to the progressive conventions of the current work-related debate about CIM (Brödner 1986). The part project CAD, by contrast, created a novelty due to its pronounced user involvement and interdisciplinary project practice. The CAD group's special ability to innovate has not only resulted in the idea of an "intelligent drawing board".* Moreover, it shows in their ability to problematize this idea again in the course of the project and as a result of

---

* This idea evolved in a Danish preliminary project, which was also characterized by a high level of user involvement.

intensive communication with designers and technical draftsmen/women. The group thus eventually arrived at the idea of the sketch board.

What then are essential conditions for achieving a truly prospective quality of R&D projects? Our experiences in this project can be summarized thus:

1. The "new" working practice in computer-aided integrated manufacturing (CAIM) must not be conceived from the point of view of the properties and possibilities inherent in computer-based/integrated production technology. Instead, a change of perspective is required. The use of work and factory scenarios is more promising of success. They are the means of both representing anticipated factory conditions and of communication among the people from the different fields involved about their specific experiences, ideas, jargons, imagination and interests. Work and factory scenarios, though containing, are not restricted by the one-dimensional formalism of technical realization. They are able to mobilize experience and to stimulate that creativity and imagination that enable people to transcend existing factories towards a higher level of human-centredness. This approach involves a degree of breaking free from the multiple ways in which one is tied to social and technical conventions. At the same time, the work (factory) scenarios are instrumental in defining technical designs, as has been shown by the Data Management Group's model (see p. 102).

2. The second dimension of a prospective development practice consists of enabling the people concerned/users to participate in the shaping of work and technology (Rauner 1988b). This ability, diametrically opposed to Taylorism and to technological determinism as it is, has traditionally been omitted from vocational (further) education. Those training concepts and instruction technologies that are geared towards adapting people to the (existing) structures of work and technology represent tremendous barriers to innovation. The idea of "enabling people to participate in the shaping of work and technology" as a guideline for the vocational training of skilled workers is not least a consequence derived from the reality of increasingly open technologies. Open architectures are to be found at all levels of computer-aided work systems – ranging from the architecture of micro-electronic components to factory architecture. These architectures need complementing by the transformation of open systems into tangible computer-aided work systems and structures that take into account the specific conditions within a given factory. The transformation of open technologies into architectures for use in companies does not require adapted qualification, but a high level of shaping ability.

3. The third dimension of a prospective development practice, finally, is shaping-oriented research. This does not conceive of the development of work and factories as being determined by technology and is, moreoever, aware of the limits, scope and restrictions inherent in individuals' views and in their ability to solve problems. Such research requires a high level of reflection, which is markedly underdeveloped in the many disciplines of academia with its pronounced division of labour. Only the productive tension between theoretical and practical competence turns science and research into factors to be reckoned with in any holistic shaping process. Sponsors' widespread and increasing pressure, even on experimental projects, to prove, prior to competition (as is the case in ESPRIT), the potential profitability of the

innovations to be developed is shortsighted and ultimately counterproductive. It goes without saying that the superiority of innovations that are part of a social modernization strategy is not amenable to evaluation by the rather limited economic methods. This, incidentally, is one of the greatest weaknesses of shaping-oriented research. As yet, nobody has succeeded in integrating into shaping-oriented research economics with its categories that necessarily abstract from the content of work and technology. This can only be accomplished by relinquishing classical economics' understanding of itself as a science.

4. Let us imagine for a moment that our project results were to be transferred to factory routine. Then we would have to deal with the part played by management, their commitment and ability to innovate. On the basis of his intimate knowledge of the company he works for, one of the engineers involved in our project stated that "another big trouble is to get traditional management to accept that individuals require a high degree of freedom in order to be flexible". Our experience of involving management in innovating companies towards a higher degree of human-centred CIM confirms that, as a rule, marketing-oriented management, due to their intellectual affinity to applied economics, is more likely to impede any long-term innovations than their engineering- and production-oriented counterparts.

References

Adorno TW, Horkheimer M (1969) Dialektik der Aufklärung. Fischer, Frankfurt
Aregger K, Frey K (1975) Ein Modell zur Integration von Theorie und Praxis in Curriculumprojekten: Das generative Leitsystem. Institut für die Pädagogik der Naturwissenschaften, Kiel
Babbage C (1832) On the economy of machinery and manufactures. Charles Knight, London
Balzert H (1986) Programmierung. Reihe Informatik 50:37–51
Bjerknes G, Ehn P, Kyng M (eds) (1987) Computers and democracy: a Scandinavian challenge. Avebury Books, Aldershot
Björn-Andersen N (ed) (1980) Artikelsamling til Socioteknisk Systemboustruktion. IFA Publications, Copenhagen
Braten S (1983) Dialogeus Vilhar: Datasamfundet. Universitetsferlaget, Oslo
Brödner P (1985) Fabrik 2000: Alternative Entwicklungspfade in die Zukunft der Fabrik. Sigma-Verlag, Berlin
Brödner P (1986) Skill based manufacturing versus unmanned factory: which is superior? Int J Ind Ergnomics 1:145–153
Burawoy M (1979) Manufacturing consent: changes in the labour process under monopoly capitalism. University of Chicago Press, Chicago
Cooley MJE (19897) Architect or bee? The human price of technology. Hogarth Press, London
Corbett JM (1987) Human work design criteria and the design process: the devil in the detail. In: Brödner P (ed) Skill based automated manufacturing. Pergamon Press, Oxford
Corbett JM (1990) Human centred advanced manufacturing systems: from rhetoric to reality. Int J Ind Ergonomics 5:83–90
Dreyfus HL, Dreyfus S (1986) Mind over machines. Basil Blackwell, Oxford
Ehn P (1988) Work oriented design of computer artifacts. Arbetslivscentrum, Stockholm
Fayol H (1949) General and industrial management. Pitman, London
Floyd C (1984) A systematic look at prototyping. In: Budde R (ed) Approaches to prototyping. Springer-Verlag, Berlin Heidelberg New York
Gordon R, Kimball LM (1985) High technology, employment and the challenges to education. Silicon Valley Research Group Working Paper No. 1. University of Santa Cruz, California

Gulick L, Urwick L (eds) (1937) Papers in the science of administration. Institute of Public Administration, Columbia University, New York

Habermas J (1981) Theorie des kommunikativen Handelns. Suhrkamp, Frankfurt/Main

Hacker W (1986) Arbeitspsychologie. Huber, Bern

Hellige HD (1984) Die gesellschaftlichen und historischen Grundlagen der Technikgestaltung als Gegenstand der Ingenieurausbildung. In: Troitsch U, König W (eds) Lernen aus der Technikgeschichte. VDI, Düsseldorf

Herzberg F (1966) Work and the nature of man. Staples Press, New York

Hildebrandt E (1987) Unternehmensplanung und Kontrollbeziehungen im Maschinenbau und Kontrolle: Eine Einführung in die Labour Process Debate. Edition Sigma, Berlin

Kling R, Scacchi W (1982) The web of computing: computer technology as social organisation. Adv Computing 21:1–93

Koslowski K (1988) Unterstützung von partizipativer Systementwicklung durch Methoden des Software Engineering. Westdeutscher Verlag, Opladen

Kuhn T (1970) The structure of scientific revolutions (2nd edn). Chicago University Press, Chicago

Lutz B (1988) Zum Verhältnis von Analyse und Gestaltung in der sozialwissenschaftlichen Technikforschung. In: Rauner F (ed) Gestalten: Eine neue gesellschaftliche Praxis. Verlag Neue Gesellschaft, Bonn

Majchrzak A (1988) The human side of factory automation. Jossey-Bass, San Francisco

Mayo E (1933) The human problems of an industrial civilisation. Macmillan, New York

McGregor D (1960) The human side of enterprise. McGraw-Hill, New York

Mertens K (1985) Steuerung rechnergeführter Fertigungssysteme. Hanser, München

Moll HH (1983) Mehr Produktivität durch weniger Arbeitsteilung. VDI Nachrichten 43:11–29

Mooney JC, Reiley AP (1931) Onward industry. Harper and Row, New York

Morgan G (1986) Images of organisation. Sage Publications, London

Noble DF (1977) America by design. Alfred Knopf, New York

Noble DF (1984) Forces of production: a social history of industrial automation. Alfred Knopf, New York

Numinen MI (1988) People or computers: three ways of looking at information systems. Studentlitteratur, Lund

Oppermann R (1983) Forschungsstand und Perspektiven partizipativer Systementwicklung. Oldenbourg, München

Perrow C (1983) The organisational context of human factors engineering. Administrative Sci Q 28:521–541

Polanyi M (1957) Personal knowledge. Routledge and Kegan Paul, London

Polanyi M (1967) The tacit dimension. Anchor Books, New York

Rauner F (ed) (1988a) Gestaltung: Eine neue gesellschaftliche Praxis. Verlag Neue Gesellschaft, Bonn

Rauner F (1988b) Die Befähigung zur (Mit) Gestaltung von Arbeit und Technik als Leitidee beruflicher Bildung. In: Heidegger G, Gerds P, Weisenbach K (eds) Gestaltung von Arbeit und Technik: Ein Ziel beruflicher Bildung. Campus Press, Frankfurt

Rauner F, Rasmussen LB, Corbett JM (1988) The social shaping of technology and work: human centred computer integrated manufacturing systems. AI Soc 2:47–61

Rosenbrock HH (1977) The future of control. Automatica 13:389–392

Rosenbrock HH (1980) Human resources and technology. In: Proceedings of the sixth world congress of the international economic association on human resources, employment and development, Mexico

Rosenbrock HH (1983) The social and engineering design of a flexible manufacturing system. In: Warman EA (ed) CAPE '83 Part 1. North-Holland, Amsterdam

Rosenbrock HH (ed) (1980) Designing human centred technology. Springer-Verlag, London

Sandberg T (1982) Work organisation and autonomous groups, Liber Forlag, Uppsala

Schon DA (1983) The reflective practitioner. Basic Books, New York

Smith A (1776) The wealth of nations. Stratton and Cadell, London

Suchman LA (1987) Plans and situated actions: the problem of human–machine communication. Cambridge University Press, New York

Taylor FW (1911) Principles of scientific management. Harper and Row, New York

Thorsrud E (1972) Worker participation in management in Norway. Institute for Labour Studies Publications, Geneva

Trist EL, Bamforth KW (1948) Some social and psychological consequences of the Longwall method of coal getting. Hum Relations 4:3–38

Ulich E (1978) Über das Prinzip der differentiellen Arbeitsgestaltung. Industrielle Organ 47:566–568

UTOPIA (1981) Report Number 1. Swedish Centre for Working Life, Stockholm

Weizenbaum J (1976) Computer power and human reason. Freeman, San Francisco

Wingert B, Duus W, Raeder M, Riehm V (1984) CAD im Maschinenbau. Springer-Verlag, Berlin

Winograd T, Flores F (1986) Understanding computers and cognition: a new foundation for design. Ablex, Norwood

Yeomans RW, Choudry A, Ten Hagen PJW (1985) Design rules for a computer integrated manufacturing system. North-Holland, Amsterdam